EAST MIDLANDS GEOTECHNICAL GROUP
THE INSTITUTION OF CIVIL ENGINEERS

Transportation geotechnics

Proceedings of the symposium
held at The Nottingham Trent University
School of Property and Construction
on 11 September 2003

Editors: M.W. Frost, I. Jefferson, E. Faragher,
T.E.J. Roff and P.R. Fleming

Thomas Telford

Organisers: The East Midlands Geotechnical Group of the Institution of Civil Engineers

Organising committee:

Dr. M.W. Frost and Dr. I. Jefferson,

School of Property and Construction, The Nottingham Trent University;

E. Faragher, and T.E.J. Roff,

Advantica Limited;

Dr. P.R. Fleming,

Department of Civil and Building Engineering, Loughborough University.

This symposium is supported by
The Highways Agency
and
The Permanent Way Institution

The Permanent Way Institution

Founded in 1884
Incorporated 1908

HIGHWAYS AGENCY

Published for the symposium organisers by Thomas Telford Publishing,
Thomas Telford Ltd, 1 Heron Quay, London E14 4JD. www.thomastelford.com

First published 2003

A catalogue record for this book is available from the British Library

ISBN: 0 7277 3249 8

Printed and bound in Great Britain by MPG Books, Bodmin, Cornwall

Preface

The East Midlands Geotechnical Group (EMGG) of the Institution of Civil Engineers was formed in 1992 with the aim of providing a learned society on geotechnical subjects, for the benefit of Civil Engineers and Engineering Geologists living in the region. EMGG activity primarily comprises the evening meetings programme, however, site visits and symposia are organised to complement this.

This symposium entitled "Transportation Geotechnics" and the associated publication is the fifth in a highly successful biennial series of such symposia run by the EMGG. The symposium aims to provide an opportunity for dissemination of current practice and discussion on recent developments in geotechnical engineering associated with transport infrastructure. Specific objectives are:

- to recognise the importance of soil mechanics as the basis of successful transportation infrastructure provision;
- to develop an understanding of a sound and holistic approach to the geotechnical engineering required;
- to present current developments across a range of issues associated with transportation geotechnics including: asset management, specifications, design and construction, and sustainability;
- to demonstrate current and innovative practice through case studies.

Of all the aspects of Civil Engineering which involve Geotechnical Engineers, arguably the largest input is in projects related to the design, planning, construction and management of the works associated with transport infrastructure. The geotechnical engineering associated with infrastructure covers all facets of the profession, from assessment, interpretation and management of potential geotechnical problems, to the development and use of state-of-the-art construction or design techniques to optimise value for money. All of this work is performed within the changing environment of the politics, standards, funding and management of the infrastructure. This symposium aims to address these issues by bringing together in one volume an eclectic collection of invited papers that highlight and address some of the current trends and innovative methods being used in transportation projects in the UK. It also considers some of the future issues that the Geotechnical Engineer must address in design, specification and construction in the context of sustainability.

The invited papers in this symposium could be grouped in a number of ways. They could be split into those associated with roads and those linked to railways. However, the geotechnical engineering of the infrastructure associated with these systems is fundamentally the same, it is only the context of the work that differs. Therefore, the papers are broadly grouped into three areas, geotechnical design and construction, management and specifications, and sustainability.

To set the scene for these papers the symposium commences with the keynote lecture "Soil Mechanics for Pavement Engineers" by 1996 Rankine lecturer, Professor Stephen Brown. His paper draws on the vast body of fundamental research he has undertaken into the behaviour of soils when subject to complex loading regimes similar to those generated by a moving vehicle. An understanding of this is vital for any Geotechnical Engineer associated with the design of the transport infrastructure.

The first group of papers on geotechnical design and construction presents reviews and case studies on some of the current areas of interest to Geotechnical Engineers. The paper by Corbet presents a state of the art review on the specification of geosynthetics. Allenby and Ropkins' paper details the construction issues associated with large jacked box tunnelling, it presents case studies where large tunnels have been preformed and inserted beneath live infrastructure to provide new routes and accesses. The Channel Tunnel Rail Link (CTRL) is the first main line railway to be constructed in this country for one hundred years, threading a 300kph railway through one of the most crowded parts of our island has required some innovative geotechnical engineering design. The paper by O'Riordan et al details the dynamic design of a piled foundation to carry the CTRL track bed over an area of marsh.

Whilst the CTRL may have a fully designed track bed to dissipate the imposed loading, the majority of our rail network carries the design remnants of its Victorian past. The paper by Armitage and Brough details innovative track bed investigation tools and explains how these can be used to predict track bed/ballast residual life, and to target maintenance. The largest maintenance problem associated with concrete pavements concerns the performance of the joints, the final paper in this section by Cudworth reviews the design and performance of concrete pavement joints and their foundations.

As the nature of funding of infrastructure schemes changes due to the introduction of private finance, the specification, management and procurement associated with such projects is also evolving. The second group of papers in the symposium considers some of these issues. Jarvis and Gilbert have written a frank and informative paper on their experiences of the design process for one of the largest private finance initiative projects to be constructed. The second paper in this group by Raybould presents a review of the asset management of earthworks that make up our transportation infrastructure. The final paper in this group, by Fleming et al, considers how recent advances in testing techniques will enable a move away from the prescriptive forms of geotechnical material specifications that are currently used.

The final group of papers investigates issues linked to sustainable construction. The paper by Ellis reviews work undertaken at the Transport Research Laboratory (TRL) to look at the use of recycled materials within infrastructure construction. If recycled or insitu materials can be used in the construction and maintenance of our infrastructure the processes will become more sustainable. The paper by Biczysko considers this by presenting data from ongoing trials on the use of stabilisation techniques to produce pavement foundations. Continuing this theme, Greenwood et al in their paper explain how "green" engineering methods may be able to help stabilise failing earthworks. The final paper by Rosenbaum and Taylor, investigates land re-use and the potential problems that may be found on disused railway land.

In organising this symposium the editors realised that it would be impossible to consider all the geotechnical methods which are of use in transportation infrastructure design. The areas this symposium has considered are the major issues facing the geotechnical engineering of the infrastructure and therefore this symposia and associated publication is both timely and of great relevance.

Finally the support for the symposium from the Highways Agency and the Permanent Way Institution is gratefully acknowledged. In addition, the editors wish to acknowledge the help of the EMGG Committee, (particularly Robin Lee), and wish to thank Nina Tyler and her team in the conference office at The Nottingham Trent University. To conclude the editors wish to express their sincere thanks to all the symposium authors for their time and considerable effort in preparing their papers.

Contents

SUSTAINABILITY

Keynote lecture: Soil mechanics for pavement engineers

Professor S.F. Brown, DSc, FREng
The University of Nottingham

Introduction

The geotechnical problems associated with the construction and maintenance of transport infrastructure embrace almost all of this important and still evolving specialist branch of civil engineering. This includes traditional stability considerations related to earthworks and retaining walls, foundation design for bridges and the shallow soil mechanics problems relating to pavement and rail track design. The expanding importance of underground construction is clear from the need for more infrastructure to be in tunnels within the urban environment. The retrofitting of services and drainage improvements around pavements, that have to remain operational, has expanded the need for trenchless technology applications.

The breadth of geotechnical engineering involved cuts across soil mechanics and rock mechanics and is rapidly expanding into the mechanics of recycled materials and industrial by-products for fill and pavement foundations. Specialist skills, such as compensation grouting to minimise the effects of underground construction on settlement of buildings, are also increasingly required.

While reinforced soil and other uses of geosynthetics are now well established techniques, they do require specialist knowledge for correct design and installation and can provide economic solutions when properly used.

It is not possible in one introductory lecture to do justice to this vast range of topics. Consequently, having sketched in the broad canvas, this contribution will concentrate on the 'Cinderella' subject of soil mechanics for pavement and rail track foundations. In doing so, it draws heavily on the Author's 1996 Rankine Lecture (Brown, 1996) and points the way to new developments and implementation of improvements in engineering practice. It also illustrates the general approach to a geotechnical problem involving a combination of theoretical modelling, laboratory and field testing, together with an application of correct design principles.

Transportation geotechnics. Thomas Telford, London, 2003

Soil mechanics for pavements

Figure 1 shows cross-sections for a number of different pavement types ranging from unsurfaced 'gravel' roads, commonly found in developing countries, to heavy duty flexible bituminous or rigid concrete pavements used for the motorway systems of industrialised countries. The figure also includes railway track as another specialist type of pavement in which the method of transmitting load to the soil differs from a highway or airport pavement but for which the essential principles of soil mechanics equally apply.

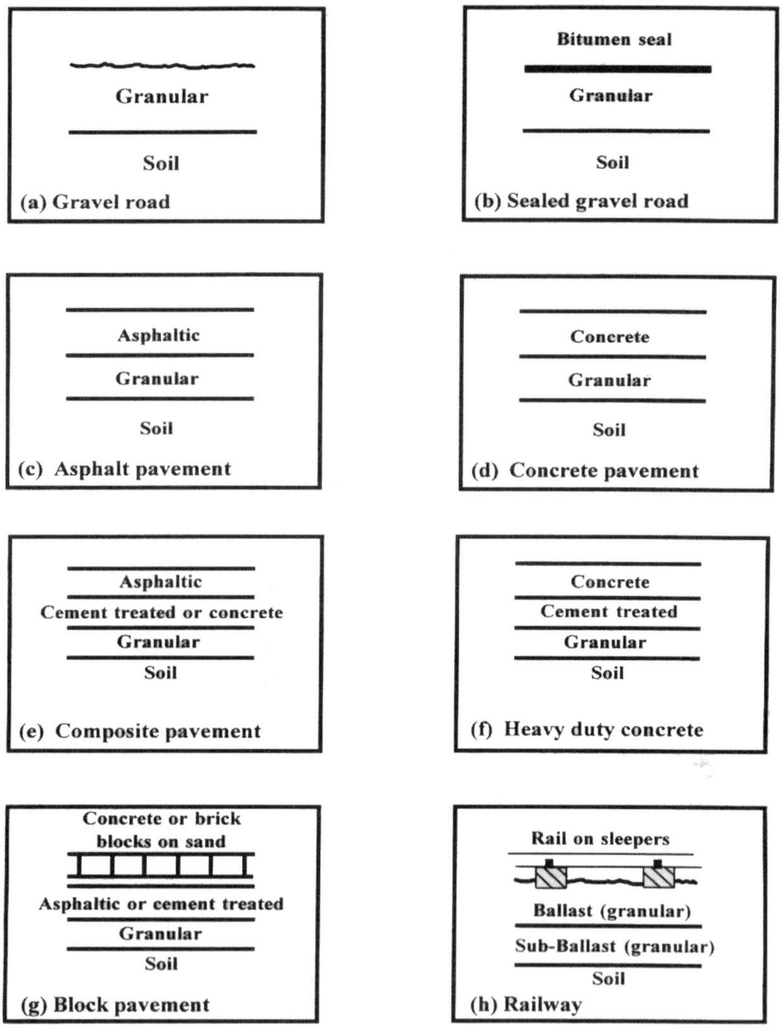

Figure 1, Variety of pavement constructions

The foundation of a pavement is defined as the sub-base layer (usually unbound granular material) and all layers below. These generally comprise the subgrade, which may be undisturbed soil in cut or recompacted soil in fill, possibly covered by a capping layer when the subgrade strength is low.

The soil mechanics requirements for pavement engineering differ significantly from those of importance in other geotechnical applications. The essential differences may be summarised as follows:

1. Soil immediately below pavements and granular materials in pavements exist above the water table but beneath a sealed surface, although this does not completely inhibit ingress of water. Both saturated and partially saturated conditions can occur, depending on the water table depth and the drainage conditions.
2. Soils and granular materials are subjected to repeated applications of stress from wheel loading. This regime has something in common with earthquake and wave loading.
3. Under a single application of a moving wheel load, a pavement responds in an essentially resilient manner. However, irrecoverable plastic and viscous strains can accumulate under repeated loading.

The magnitude of traffic-induced repeated stress experienced at formation level in a completed pavement is well below its shear strength but the number of stress repetitions in the design life will be very large. By contrast, when construction traffic is operating at sub-base level in a partially completed pavement, the stress levels are much higher but the number of repetitions is much smaller. Figure 2 illustrates both situations with vertical stress measurements taken at two experimental sites. Data from a haul road experiment on a soft clay site is shown in Figure 2(a), (after Little,1993). The measurements were taken below a 300mm granular layer loaded by a truck with an 80kN axle load. The undrained shear strength of the soil varied between 75 and 80kPa. The example in Figure 2(b) is taken from a colliery spoil haul road where the construction involved 210mm of asphalt over a 300 mm sub-base (Brunton and Akroyde, 1980). The subgrade had a higher shear strength but the stress levels are seen to be very much lower than for the unsurfaced case.

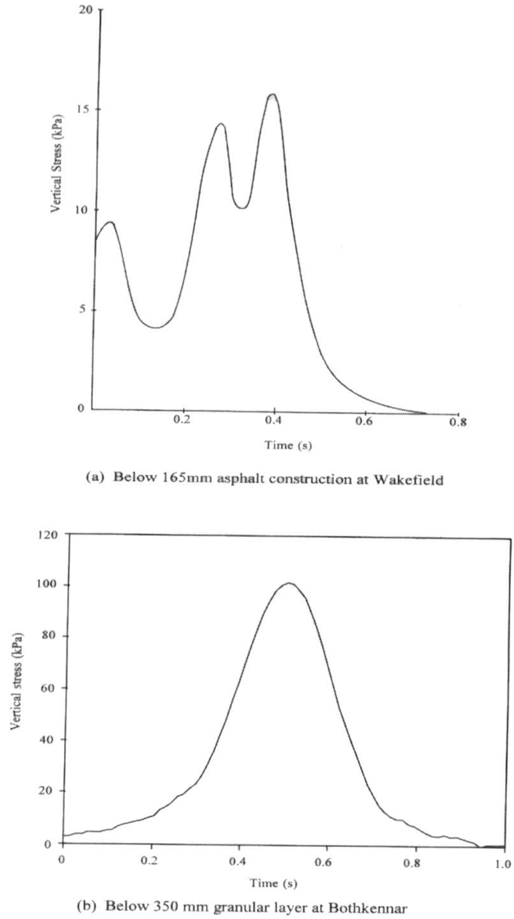

Figure 2, Field measurements of stress

Pavement design and analysis

While much pavement design still uses essentially empirical methods, techniques based on the use of theoretical concepts are increasingly adopted, particularly where heavy loading or unusual conditions are involved. The commonest approach is to use linear elastic analysis of the pavement as a layered system. For design, the computed tensile strain at the bottom of the asphalt layer should not exceed a particular value, in order to prevent fatigue cracking, and the vertical strain at formation level is limited to reduce the possibility of rutting in the wheel tracks. The allowable values of strain vary with location and local practice with respect to material types. Other criteria are appropriate when cement treated materials are used. For heavy-duty pavements in the UK, the large thickness of bound material is now considered to eliminate the possibility of cracking from the bottom of the layer but such pavements can experience surface-initiated cracks (Nunn, 1997).

There are many assumptions in the simplistic approach outlined above and these are discussed elsewhere (Brown, 1997). For the purposes of this paper, comment is focussed on the issues relating to the properties of the granular layer(s) and the soil.

The non-linear stress-strain characteristics of soils are now well recognised in geotechnical engineering as important for modelling low strain problems, which include pavements. Although the assumption that a pavement will respond in an essentially resilient manner under a single load application suggests the use of elastic theory, non-linear effects can be significant in many cases. Figure 3 illustrates the stress-strain relationship for a soil under a single load cycle at a relatively low stress level. While some irrecoverable strain is apparent, the resilient (recoverable) strain is of much larger magnitude but the relationship is clearly non-linear. The problem of non-linearity can be dealt with either by using an iterative linear elastic approach or through finite element analysis but the accumulation of permanent strains (plastic or viscous), which cause rutting, is a more complex problem.

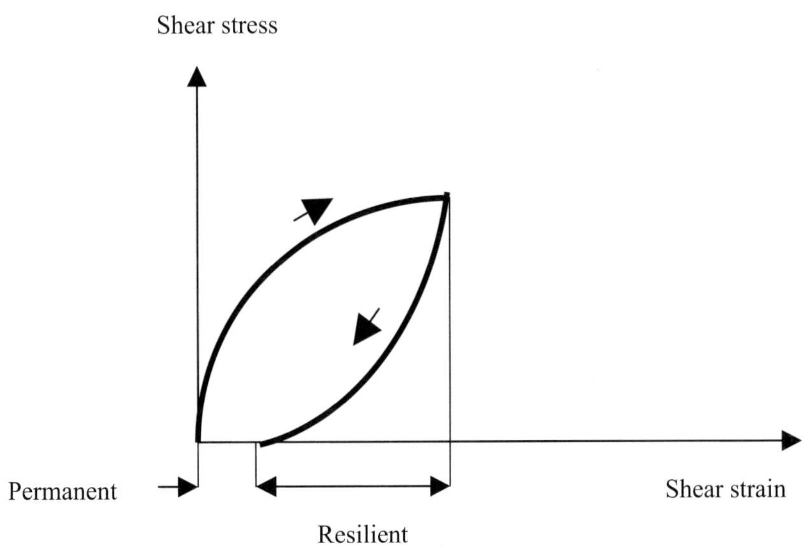

Figure 3, Resilient and permanent strains

Recognising the difficulty of reproducing field conditions in a laboratory test, the data in Figure 4 are of interest in illustrating insitu non-linearity for a compacted silty-clay. The data were drawn from insitu measurements of stress and strain resulting from dynamic plate loading tests (Brown and Bush,1998). Similar data were generated for crushed rock (Brown and Pell, 1967).

If some form of elastic analysis is to be used for pavements, then assumptions of linear behaviour are safest when the influence of the non-linear lower layers is least. This will be the case for heavy-duty pavements where the response of the bound layers will dominate. For pavements with thin surfacings, however, non-linear analysis will be required.

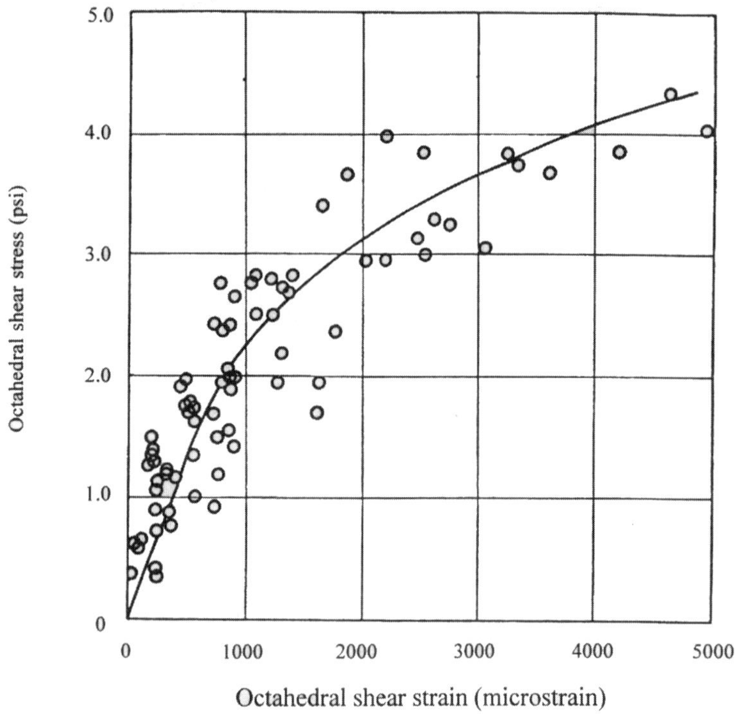

Figure 4, Non-linear insitu response of a silty clay

A sensitivity analysis reported by Dawson and Plaistow (1997) based on finite element computations revealed that a change in the resilient modulus of the subgrade from 40 to 90 MPa caused a change in the maximum asphalt tensile strain of less than 2%. Changes in the resilient characteristics of the granular layer over a realistic range, using non-linear models, were more significant, causing the asphalt tensile strain to vary by up to 70%. It follows that the correct characterisation of the granular layer is of more importance for pavement design than that of the subgrade.

The above guidelines apply for design analysis but not for the back-analysis involved in structural evaluation of pavements with data from the Falling Weight Deflectometer (FWD) (Figure 5). This equipment allows the surface deflection profile of a pavement to be determined with high accuracy and the process of back-analysis involves computation of the effective modulus of elasticity for each principal layer to provide a match between theory and measurement.

Figure 5, Falling (Heavy) Weight Deflectometer (FWD)

Figure 6 shows a typical vertical variation in vertical resilient strain computed for an asphalt pavement structure. Surface deflection is the summation of strain with depth and the figure clearly shows, even for a pavement with an asphalt base, that the major contribution comes from the subgrade. Consequently, the characteristics of this layer must be properly modelled for realistic back-analysis. Hence, the non-linear properties of the soil have to be accounted for. Further details are described by Brown et al (1993) .

Wheel track rutting arises through the accumulation of vertical permanent strains which can, in principle, include contributions from all layers. For thick asphalt pavements, rutting usually arises from permanent deformation in the bituminous layers, often the surface course. For pavements with thin bituminous layers, the granular layer and subgrade are likely to dominate, particularly if drainage conditions are unsatisfactory. For construction traffic operating on the pavement foundation, rutting is a major concern and must be limited to avoid undue damage to this layer or to the subgrade below.

In addition to quantifying the non-linear resilient characteristics of granular materials and soils, it is clearly necessary to understand the relationship between the accumulation of plastic strain and applied stress together with relevant variables that may influence this relationship.

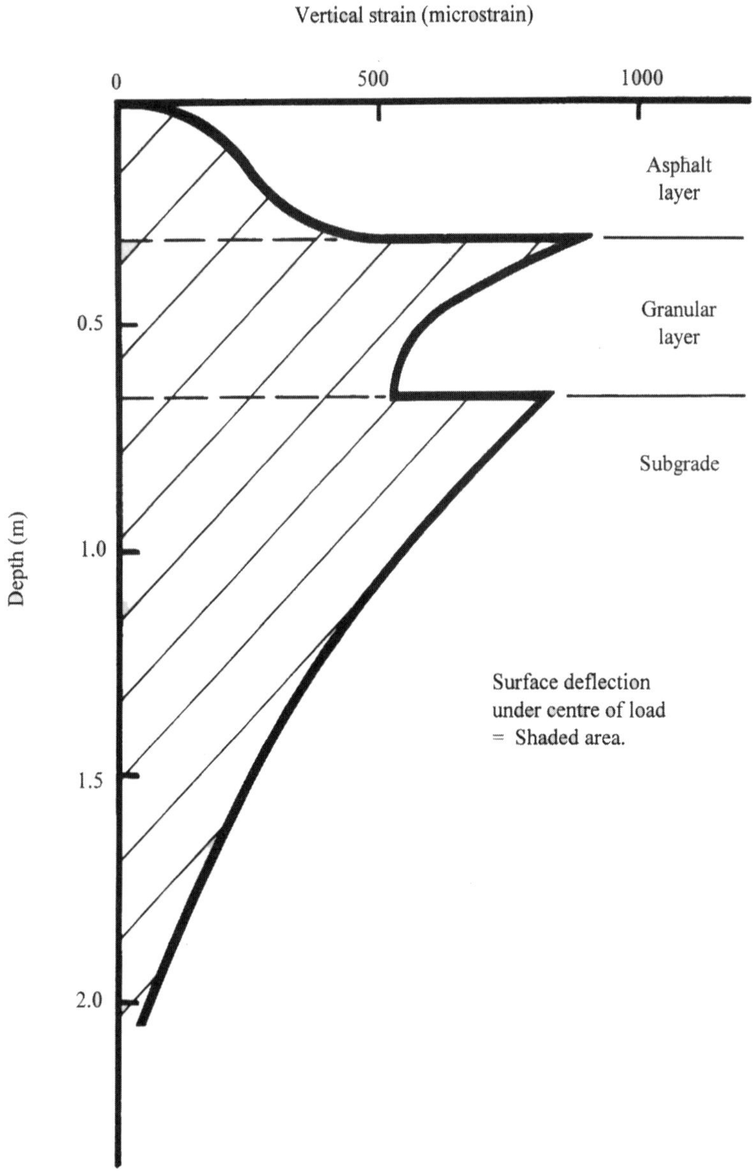

Figure 6, Variation of vertical resilient strain through an asphalt pavement

Failure mechanisms in rail track are well described by Selig and Waters (1994). Under repeated loading, differential permanent strains develop in the ballast which cause the rail line and level to change and the ride quality to deteriorate.

Insitu stress conditions

Equilibrium stresses

The response of an element of soil to applied load depends crucially on its consolidation stress history and the current effective stress state. Since formation levels in subgrades exist above the water table, the determination of pore pressure and, hence, effective stress, is generally not straightforward. Immediately above the water table, where the soil is saturated, the negative pore pressure is proportional to height above the water table. The proportionality breaks down as the soil becomes partially saturated at greater heights. For fine-grained soils and shallow water tables, conditions that apply to most of the UK, the situation is simplified since saturation conditions may be assumed up to formation level, certainly for design purposes, and pore pressure can be determined.

The soil beneath a pavement may be in its natural undisturbed state or be remoulded depending on whether the section of pavement is in a 'cut' or a 'fill' area. These two situations require separate consideration. For saturated undisturbed clays, the stress history is represented by Figure 7. The parameters used in this figure are:

Mean normal effective stress, $p' = (\sigma'_v + 2\sigma'_h)/3$ (1)

Deviator stress, $q = \sigma'_v - \sigma'_h$ (2)

where σ'_v and σ'_h are the vertical and horizontal effective stresses, respectively.

Specific volume, $v = 1 + w\,G_s = 1 + e$ (3)

where w = water content,
G_s = specific gravity of soil solids
e = void ratio.

Figure 7 shows pre-consolidation involving compression to point C and subsequent swelling to point A, all under anisotropic conditions (zero lateral strains). This historical sequence generates an over-consolidated soil, being typical of a stiff clay deposit.

The construction operation involves three processes which will influence the effective stress in the soil. These are:

1. Removal of overburden during earthworks construction.
2. Lowering of the water table by provision of drainage.
3. Addition of overburden due to the pavement construction.

The nett effect of these operations will be for the effective stress state to move from point A in Figure 7 to point P via P'.

For soil which is cut, transported and compacted as fill in an embankment, the effective stress regime is rather different and less well understood. Brown[1] suggested that the nett effect of these operations is to significantly reduce the overconsolidation ratio.

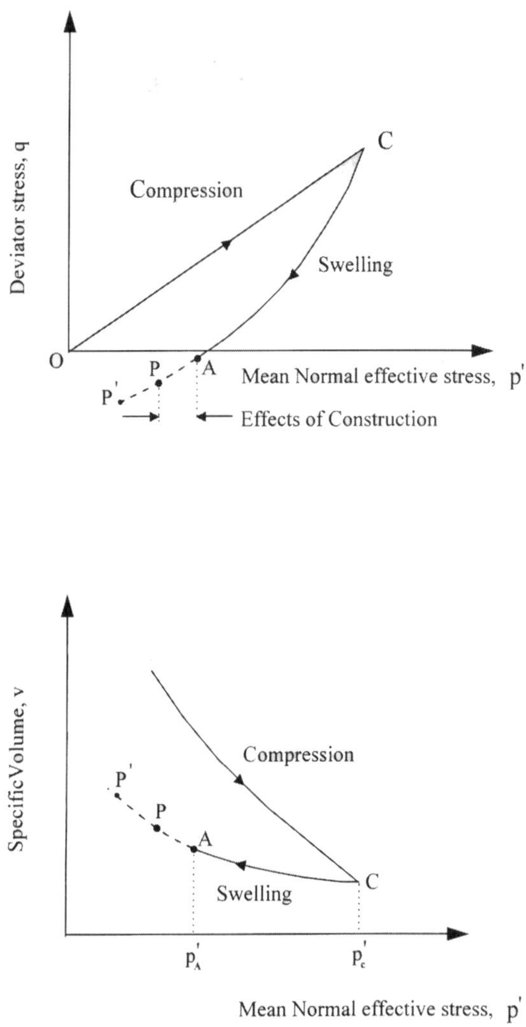

Figure 7, Stress history for a stiff clay subgrade

Traffic Stresses

A moving wheel load will result in a transient stress pulse being transmitted to the soil element. This might involve a change in total stress along a path such as PT, in Figure 8 for material at P in the cutting. Since the loading event is rapid, there will be no change in effective stress so a transient pore pressure, δu, develops at peak stress. The effective stress follows the path PEP, corresponding to the transient deviator stress of magnitude q_r.

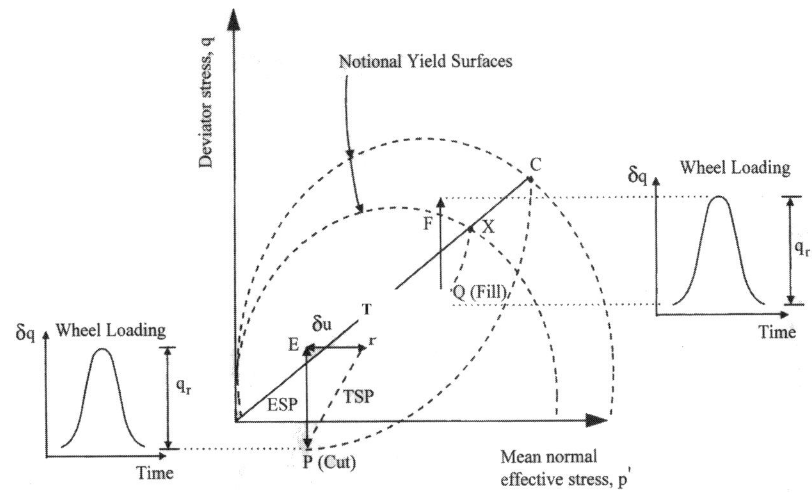

Figure 8, Traffic included stress paths

Soils are essentially elasto-plastic materials. Hence, the stress history for soil at P will have established a yield surface through C, the preconsolidation stress, so that in the zone beneath this yield surface it should not develop plastic strains. Hence, the stress path PEP should result in soil behaving as a resilient material.

The soil on the embankment at point Q in Figure 8 is much nearer to its associated yield surface. Hence, the same traffic induced stress, q_r, will cause the effective stress to probe beyond the yield surface at F, resulting in some plastic strains developing.

Laboratory testing has shown that plastic strains can accumulate even within the static yield envelope. McDowell and Hau, (2002) have adopted the three surface kinematic hardening model developed by Stallebrass and Taylor (1997) and extended it to accommodate cyclic loading. The essential feature of this approach is that it incorporates an inner yield surface in the form of a 'bubble' around the current stress state as shown in Figure 9 based on Al-Tabbaa (1987). This reproduces the initial high stiffness on loading or unloading and allows some plastic strain to accumulate within the static yield envelope.

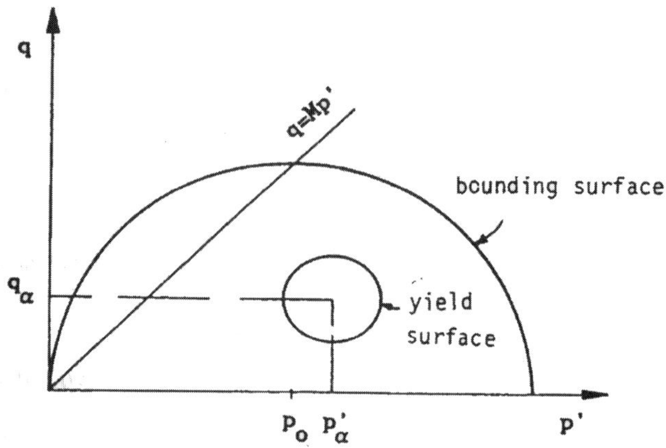

Figure 9, "Bubble" model (after Al-Tabbaa, 1987)

Laboratory testing

Resilient characteristics

Figure 10 illustrates the general stress regime experienced by an element of material in or below a pavement structure as a result of a moving wheel load within the plane of the wheel track, ie the longitudinal plane. There are pulses of vertical and horizontal stress accompanied by a double pulse of shear stress with a sign reversal on the vertical and horizontal planes. Figure 11 shows the associated pattern of principal stresses illustrating the rotation of principal planes which takes place.

One approach to laboratory testing is to select the equipment which reproduces the field situation. Clearly, for pavements this would demand complex facilities. A close match to field conditions can be obtained by use of a Hollow Cylinder Apparatus (HCA). This allows control of both normal and shear stresses in a manner which can reproduce the insitu regime, as shown in Figure 12, and is a more accurate approach than use of the Simple Shear Apparatus, where accurate control of the stresses and measurement of deformations are difficult.

For testing granular materials with particle sizes up to 40mm, a larger repeated load triaxial apparatus has been developed catering for 150mm diameter specimens (Brown et al, 1989). A feature of all these pieces of equipment is that they incorporate 'on-specimen' deformation measuring transducers to ensure sufficient resolution of the small strains involved for defining resilient response. Simpler versions, utilising pneumatic actuators and displacement transducers that can easily be clipped on to the specimen, have also been developed for routine use.

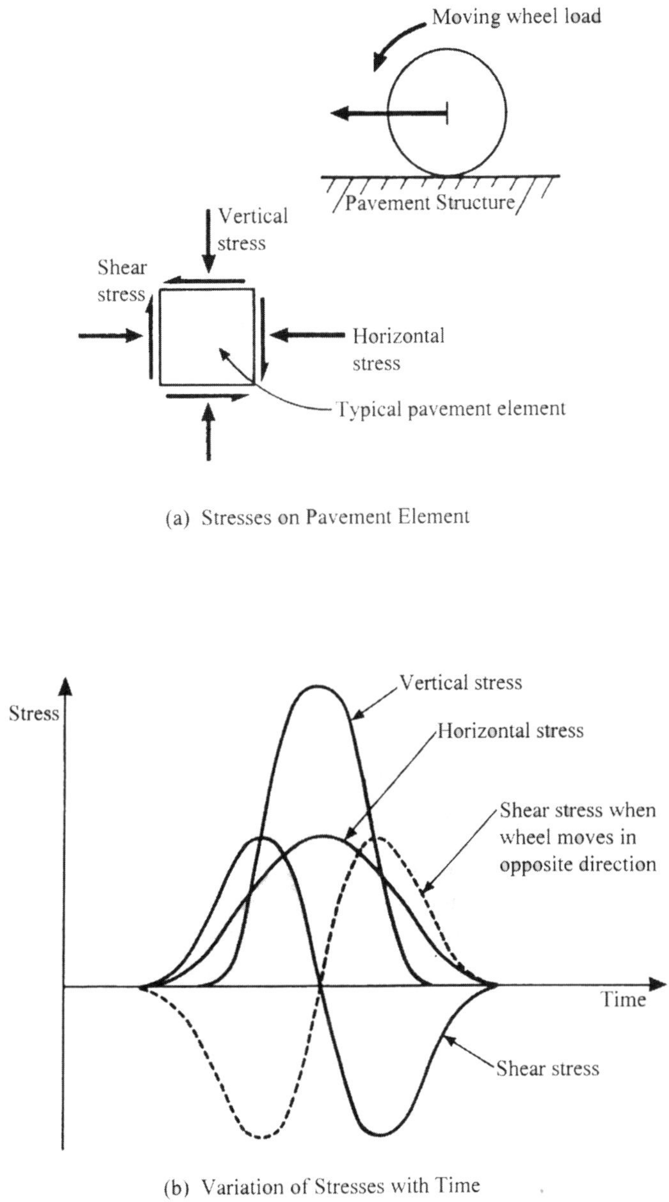

Figure 10, Stress regime under a moving wheel load

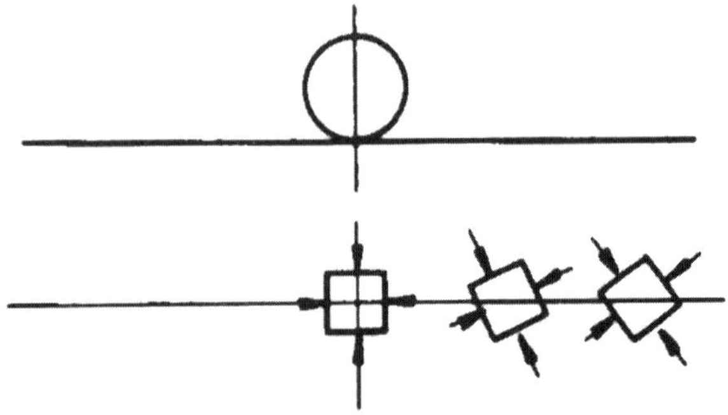

Figure 11, Rotating principal planes

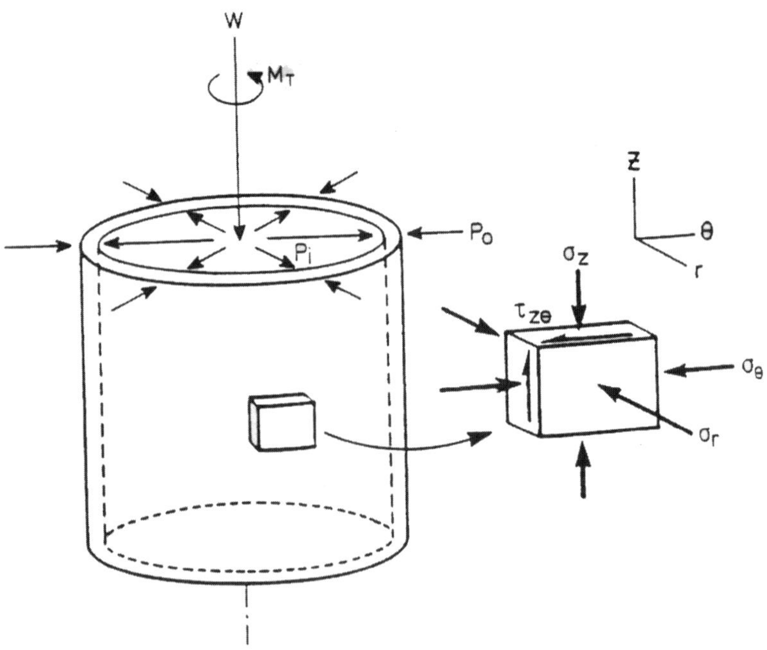

Figure 12, Hollow cylinder test

Much more research has been devoted to the measurement of resilient soil properties under repeated loading than to the accumulation of plastic strains. Brown et al, (1987) developed the following empirical model for a silty clay:

$$G_r = q_r\{p'_o / q_r\}^m / C \qquad (4)$$

where C and m depend on the soil type.

An important feature of this relationship is that it emphasises the importance of the stress ratio (q_r/p'_o) and reflects the non-linear resilient behaviour. Its main shortcoming is that unrealistically high values of resilient modulus are predicted at low deviator stresses. Hence, when applied to pavement analysis, it predicts an increase in stiffness with depth but the values become too high at large depths (Brown et al, 1992).

In other branches of geotechnical engineering, stress-strain non-linearity is expressed in terms of a relationship between a normalised shear modulus and shear strain. The normalising parameter is the value of shear modulus (G_o) at very low strains, which is the maximum value achieved. Roblee et al (1994) published the relationships shown in Figure 14, drawn from cyclic loading tests largely associated with earthquake related research. The relationship between G/G_o and cyclic shear strain is shown to depend on the Plasticity Index of the soil. Various proposals have been made for estimating G_o for clays, which typically involve overconsolidation ratio and plasticity index and this approach could usefully be adopted for pavement analysis (Brown, 1996).

The resilient behaviour of granular material has been shown to be of prime importance for pavement analysis, both in connection with foundation design and for completed construction. For these materials, both volumetric and shear strains need to be quantified since volume change will generally occur under undrained repeated loading unless the degree of saturation is very high. Most of the research used to define the non-linear stress–resilient strain behaviour of granular materials has used material in the dry state. Repeated load triaxial tests reported by Pappin and Brown (1980) generated the data for defining stress-dependent bulk and shear moduli. Their results showed that the resilient shear modulus is principally a function of the stress ratio (q_r/p'_r) and that the bulk modulus depends mainly on p'.

Application of these relationships through finite element analysis by Brown and Pappin, (1985), allowed derivation of an equivalent representative value of Young's modulus for a typical 'type 1' sub-base of 100MPa for use in linear elastic analysis of pavements with bound bases.

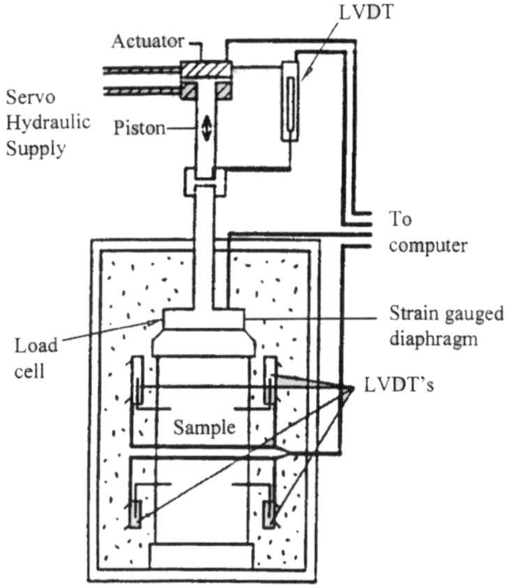

(a) Axial stress and deformation system

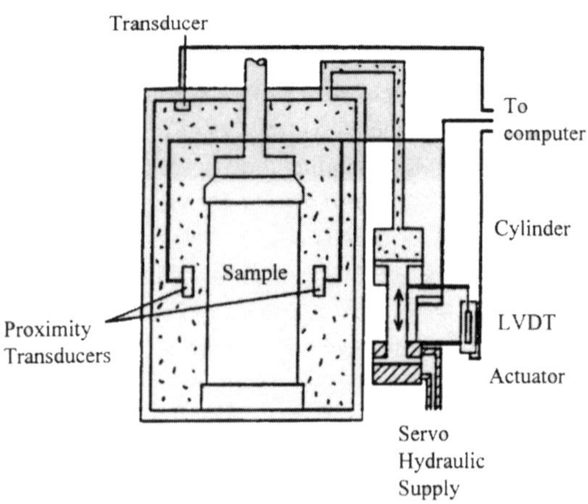

(b) Confining stress and radial deformation system

Figure 13, Repeated load triaxial test

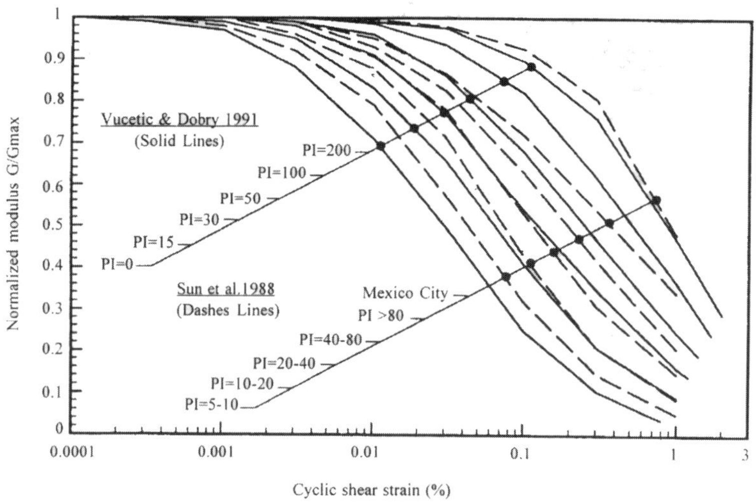

Figure 14, Non-linear relationships (after Roblee et al,1994)

Plastic strains

The detailed prediction of rut development based on correct elasto-plastic analysis of the lower layers remains a challenge. However, an approach of interest for design is that based on the concept of ‘shakedown’ or ‘threshold stress’. Repeated load triaxial testing of soils has shown that a threshold level of deviator stress can sometimes be defined, above which serious plastic strains accumulate and below which the strains and pore pressures are negligible. Figure 15 illustrates this point for two specimens of a saturated silty clay subjected to successive bursts of repeated loading at gradually increased deviator stress levels (Brown et al, 1987). A similar pattern emerged from data obtained by Loach (1987) on compacted specimens of three clays with degrees of saturation in excess of 85%. These data are summarised in Figure 16, which shows a relationship between the threshold level of deviator stress, and soil suction that could be of practical use for design, (Brown and Dawson, 1992).

More recently, the structural engineering concept of 'Shakedown' has been applied to pavement engineering following a suggestion by Sharp and Booker, (1984). The essential principle is the same as that of the threshold concept but it has a sounder theoretical basis. During early load cycles, some plastic strain will accumulate but, if this is small and the soil then settles to an equilibrium state with no further accumulation, it is said to have reached ‘shakedown’. The maximum repeated shear stress at which this can occur is known as the ‘shakedown limit’. A useful summary of this approach has been presented by Werkmeister et al (2001). Figure 17 shows a stress-strain relationship for soil below the shakedown limit.

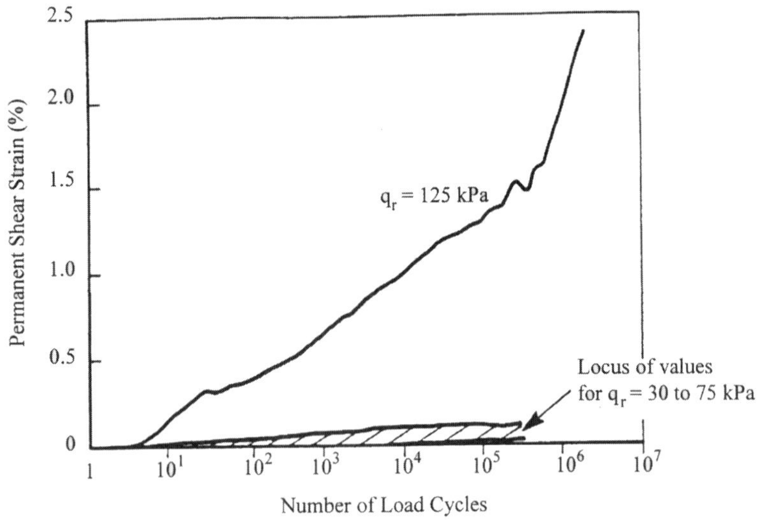

Figure 15, Repeated load triaxial data for a stiff clay

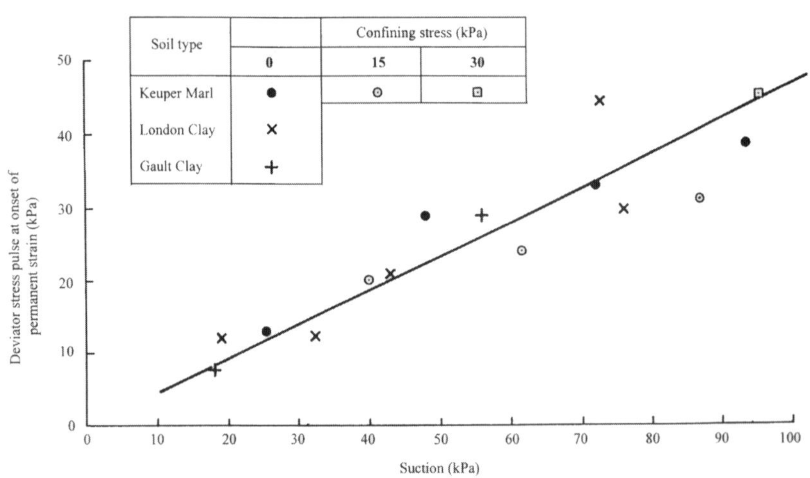

Figure 16, Threshold stress as a function of suction

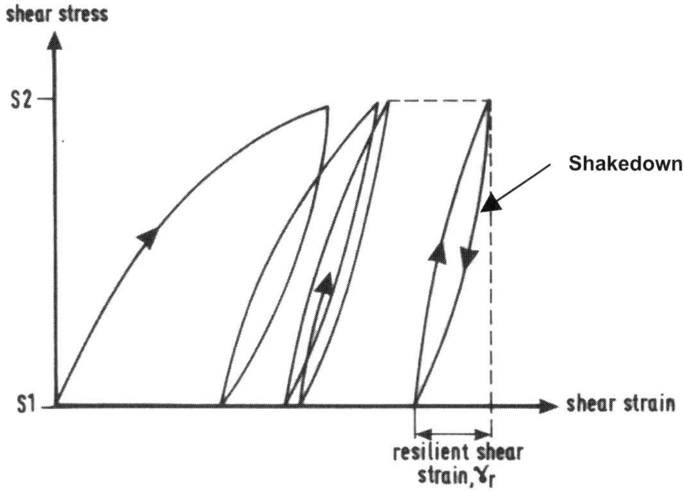

Figure 17, Shakedown condition

California Bearing Ratio

The Californian Bearing Ratio (CBR) is still extensively used as the means for characterising subgrades for pavement design purposes. For current UK design practice (DMRB, 1994), foundation layer thicknesses are empirically determined from simple charts based on subgrade CBR. This may be estimated from plasticity data, approximate water table position, 'construction conditions' and choice of a 'thick' or 'thin' pavement An analytical approach may be used as an alternative and a 'stiffness modulus' for the subgrade can be estimated from the CBR using an empirical equation reported by Powell et al (1984), viz:

$$E_r = 17.6\ \text{CBR}^{\,0.64}\ (\text{MPa.}) \qquad (5)$$

In view of the fact that CBR is essentially an indirect method for measuring undrained shear strength, it is surprising that it should be used to estimate what is, effectively, the resilient modulus (E_r) of the soil. Brown et al (1990) demonstrated that E_r was not a simple function of CBR but depended on soil type and the applied deviator stress level. Sweere (1990) could find no correlation between the CBR and resilient modulus for a range of granular materials.

While much of UK current practice using the CBR concept is based on research at TRL, Croney (1977), who lead this work noted: *'The shear strength of soil is not of direct interest to the road engineer in connection with the behaviour of pavements under traffic. To provide a satisfactory subgrade, the soil should operate at stress levels within the elastic range. The pavement engineer is, therefore, more concerned with the elastic modulus of the soil and the behaviour under repeated loading'.*

Field testing

Laboratory testing of small elements always raises questions about whether the results are representative of field conditions for the soil in bulk. Field testing, though more expensive, has therefore become an increasingly important part of geotechnical engineering.

Work in recent years has concluded that a dynamic plate-loading test is appropriate for assessing the resilient modulus of granular layers. This test is a simplified version of the more sophisticated Falling Weight Deflectometer apparatus (Figure 5), now used extensively for the structural evaluation of pavements and which can also be applied to rail track (Figure 18).

Figure 18, The Falling Weight Deflectometer in use on rail track

In both cases, a load pulse is generated by a mass falling onto a spring above a load platen. The peak load is measured and the corresponding resilient deflection of the load platen or, more accurately, the material below it via a hole in the centre of the load platen, is recorded. An effective foundation stiffness modulus (E_f) can be computed using classical Bousinesq theory, adjusted for the rigid platen, as follows:

$$E_f = \frac{P(1-v^2)}{2d\,a} \qquad (6)$$

where: P = Applied load
a = Plate radius
ν = Poisson's ratio
d = Measured plate deflection

A formal standardised procedure for foundation testing is being developed for the Highways Agency. Acceptable values for equivalent foundation stiffness are different for the two pieces of apparatus, reflecting the different levels of stress which each is able to apply and the consequent effects of non-linearity. This technique was used to compare two sections of the A50 Derby Southern By-Pass, tested at various stages of construction. One section was in cut and the other on an embankment. The soil was a silty-clay in both cases, the capping was locally won sand and gravel, the sub-base was crushed limestone in the cutting and granite on the embankment and the road base was a dense bituminous material. The FWD was used to obtain deflection data and the results were back-analysed to determine effective resilient moduli for pavement layers as construction proceeded.

The data in Table 1, for the section in cut, shows how the capping and subgrade mobilised higher resilient moduli when covered by sub-base and road base due to increased confining stress and decreased deviator stress in keeping with the non-linear stress-strain relationships.

Table 2 shows values of equivalent foundation stiffness computed using equation (6). The values increased as tests were conducted successively on the subgrade, capping and sub-base of the embankment. This shows how the effective formulation modulus increases as additional layers are added.

Table 1, Back analysed effective values of resilient modulus for road in cutting

	Effective E_r (MPa)	
Pavement Layer	**Test on Capping**	**Test on Road Base**
Road base	-	3200
Sub-base	-	240
Capping	90	200
Subgrade	70	200

Table 2, Equivalent foundation stiffness values for road on embankment

Test on:	**Equiv. found. Stiffness (MPa)**
Subgrade	30
Capping	50
Sub-base	90

Concluding remarks

This review of soil mechanics principles for pavement foundations illustrates just one area of geotechnical engineering related to transport infrastructure. The approach taken is considered typical of that needed in general, since it includes various elements aimed at obtaining improved design procedures. These embrace theoretical modelling, design criteria, laboratory testing and field testing.

The application of soil mechanics principles to pavement engineering is of most importance for constructions with thin surfacings but the discussion has illustrated two areas where applications to heavy duty pavements are also desirable.

- Correct modelling of the sub-base for pavement design
- Correct modelling of the subgrade for back-analysis of surface deflection.

In both areas it is the non-linearity of stress-strain relationships under resilient conditions that need to be accounted for.

A perspective on current, mainly empirical, practice which involves use of the CBR as a means of characterising subgrades reveals it to be an inadequate parameter for determination of resilient properties.

Progress has been made towards a performance-based specification for pavement foundations involving field testing to assess an effective resilient modulus. Details are expected to be published in the near future.

References

Al-Tabbaa A, *Permeability and stress-strain response of speswhite kaolin*, Ph.D thesis, University of Cambridge, 1987.

Brown S F, 36th Rankine Lecture: *Soil Mechanics in Pavement Engineering*, Géotechnique, Vol. 46, No. 3, pp 383-426 1996.

Brown S F, *Achievements and challenges in asphalt pavement engineering*, Keynote Lecture, Proc. 8th Int. Conf. on Asphalt Pavements, Vol. 3, pp 19-41, Seattle, 1997.

Brown S F and Pell P S, *An experimental investigation of the stresses, strains and deflections in a layered pavement structure subjected to dynamic loads*, Proc. 2nd Int. Conf. on the Struct. Design of Asphalt Pavements, Ann Arbour, Michigan, pp 487-504, 1967.

Brown S F and Pappin, J W, *Modelling of granular materials in pavements*, Transp. Res. Rec. 1022, pp 45-51, 1985.

Brown S F, Loach S C and O'Reilly M P, *Repeated loading of fine grained soils*, Contractor Report 72, Transp. Res. Laboratory, 1987.

Brown S F, Tam W S and Brunton J M, *Structural evaluation and overlay design: analysis and implementation*, Proc. 6th Int. Conf. On the Struct. Design of Asphalt Pavements, 1, Ann Arbour, Mich., pp 1013-1028, 1987.

Brown S F, O'Reilly M P and Pappin J W, A repeated load triaxial apparatus for granular materials, Unbound Aggregates in Roads, Butterworths, pp 143-158, 1989.

Brown S F, O'Reilly MP and Loach SC, *The relationship between California Bearing Ratio and elastic stiffness for compacted clays*, Ground Engineering, 23, No 8, pp 27-31, 1990.

Brown S F and Dawson A R, *Two-stage approach to asphalt pavement design*, Proc. 7th Int. Conf on Asphalt Pavements, 1, Nottingham, 16-34, 1992.

Brown S F and Bush D I, *Dynamic response of model pavement structure*, Transportation Eng. Journ. Of ASCE, TE4, pp 1005-1022, 1998.

Brunton J and Akroyde P M, *Monitoring the performance of a full-scale experimental pavement*, Proc. 3rd Int. Conf. on Bearing Capacity of Roads & Airfields, Trondheim, pp 585-594, 1980

Croney D, *The design and performance of road pavements*, London, HMSO, 1977

Dawson A R and Plaistow L, *Parametric study – flexible pavements*, Flexible Pavements, Balkema, pp 229-237, 1993.

Design manual for roads and bridges (DMRB): Vol 7, *Pavemetn design and maintenance*, Highways Agency HMSO, 1994.

Little P H, *The design of unsurfaced roads using geosynthetics*, Ph.D thesis, University of Nottingham, 1993

Loach S C, *Repeated loading of fine grained soils for pavement design*, Ph.D thesis, University of Nottingham, 1987.

McDowell G R, and Hau K W, *A simple non-associated three surface kinematic hardening model*, Geotechnique 53, No 4, pp 433-437,

Nunn M, *Long life flexible roads*, Proc. 8th Int. Conf. On Asphalt Pavements, Vol. 1, Seattle, pp 3-16, 1997.

Pappin JW and Brown S F, *Resilient stress-strain behaviour of a crushed rock*, Proc. Int. Symp. On soils under cyclic and transient loading, 1, pp 169-177, Swansea, 1980.

Powell W D, Potter J F, Mayhew H C and Nunn M E, *The structural design of bituminous roads*, Lab. Report 1132, TRRL, 1984.

Roblee C J, Xiang-Song L, Chan C K, Idriss I M, Wang G, Herrmann L R and Jackura K A, *Feasibility of a tool for in situ measurements of material properties of clays over a wide strain range*, Dynamic Geot. Testing II, ASTM STP 1213, pp 134-161, 1994.

Selig E T and Waters J M, *Track geotechnology and substructure management*, T Telford, London, 1994.

Sharp R and Booker J, *Shakedown of pavements under moving surface loads*, pp 1-14, ASCE Journal of Transport Engineering, No 1, 1984.

Stallebrass S E and Taylor R N, *The development and evaluation of a constitutive model for the prediction of ground movements in overconsolidated clay*, Geotechnique 47, No 2, pp 235-253, 1997.

Sweere, G.T.H. *Unbound Granular Bases for Roads*. PH.D Thesis, Delft University of Technology, Delft, The Netherlands. (1990):

Werkmeister S, Dawson A R and Wellner F, *Permanent deformation behaviour of granular materials and the shakedown theory*, Transportation Research Record No 1757, pp 75-81, 2001.

Use of geosynthetics in transportation projects

S. Corbet
FaberMaunsell Ltd.

Introduction

Geosynthetics have been used in highways and other transportation projects in the United Kingdom since the early 1970's. The application of geosynthetics to geotechnical and hydraulic engineering has allowed engineers to make savings in the uses of natural materials and in many cases significant financial savings have resulted. In considering highway widening geosynthetics have allowed projects to be realised without the need to purchase additional land. In this paper some examples of the uses of geosynthetics in UK Transportation projects are presented.

What are geosynthetics

Geosynthetics are sheet products which comprise the following generic families of products defined below, for a detailed definition of each term the EN standard EN 10318 (1) provides details :

- Geotextiles
- Geomembranes (or Geosynthetic Barriers in the new Euro terminology)
- Geogrids
- Geonets, Geomats
- Geocells, Geostrips
- Geospacers
- Geofoam

Common polymers used for geotextiles are :

- Polypropylene (PP)
- Polyethylene (PE)
- Polyamide or Nylon (PA)
- Polyester (PET)
- Polymers PP, PE and PA are normally used for filters, separators and protection
- PET is normally stronger and stiffer and is often used in reinforcement geotextiles.

Geogrids are normally manufactured from :

- Polyester , normally covered with PVC or polyethylene to protect the fibres from mechanical damage and chemical attack
- Polyethylene or Polypropylene, drawn grids, either uniaxially or biaxially drawn
- Welded strips formed from drawn high modulus polypropylene or bundles of polyester fibres in a PVC or polypropylene sheath.

Geosynthetic Barriers can be produced from any of the polymers

- High density PE (HDPE)
- Linear low density PE (LLDPE)
- Ethylene copolymers (ECP)
- Flexible polypropylene (FPP)
- Plasticised polyvinyl chloride (PVCp)
- Others : bituminised , butyl rubber, GCL's

Geotextiles are often described by their manner of production. The way geotextiles are made imparts many of the characteristics which are important in the design process. Normal methods are:

- Weaving, tapes or fibres (single or bundles)
- Non-woven - spun heat bonded filaments or needle punched filaments or fibres
- Knitted, very rare used for very special products

Geogrids, there are 3 basic types of geogrid:

- Punched or extruded, drawn sheets (Tensar or Tenax) with a high resistance to site damage, but relatively large extensions at failure.
- Coated PET filaments (Fortrac), low strain at failure, but can be damaged.
- Polyethylene coated PET : can be made very strong, good resistance to damage at high strengths.

Functions of geosynthetics

- Separation
- Filtration
- Drainage
- Reinforcement
- Protection
- Surface Erosion Control
- Barrier layer

Typical uses of geosynthetics in transportation projects

Given the wide range of geosynthetic products available there are many uses of geosynthetic. The specific uses are almost only limited by the imagination of the designer, however some typical uses of geosynthetics and the benefits are described in Table 1

Table 1: Uses of geosynthetics in transportation

Description & Function	Products Normally used	Benefit
Separation / Reinforcement under temporary site roads	GTX	Road has less ruts, less fill used to construct the road for a given life
Filter around drains	GTX	Reduces clogging of drain pipes, can eliminate the need for graded filters
Geocomposite drains edge of carriageway or as structural drainage	GTX & Geomesh or Geospacers	Eliminates the need for imported drainage media, can be installed in very narrow trenches or with trenching machines
Band drains – to increase rate of dissipation of pore water pressures	GTX NW & Geomesh	Increased rate of pore water dissipation and quicker settlement or works
Reinforcement at base of embankment fill	GTX W or Geogrid	Prevents lateral sliding of embankments and enhances stability for deep seated failure planes
Reinforcement in over steep slopes	Geogrid or GTX W	Allows fills to be placed at inclinations steeper than angle of internal friction and reduces the volume of fill
Reinforcement in RE Soil walls and Modular blockwork walls	Geogrid or GTX W	Minimises amount of concrete required in retaining wall construction, allows use of low grade fills
Surface erosion protection on slopes	Geomesh or Geoweb	Reduces amount of soil washed off slopes and helps vegetation become established
Barriers in run off ponds or SUDS	Geomembranes or Barriers	Prevents contaminates in run-off entering soil or ground water
Protection to barriers in run off ponds	GTX NP	Prevents or reduces risk of puncture, by stones etc.
Ultra lightweight fill to reduce foundation loadings	Geofoam	Reduces loads on foundation soils or services beneath foundations
Barriers to separate the construction from underlying contamination and gases	Geomembranes, Geospacers	Allows construction to proceed without removal of contaminated material
Separation beneath railway ballast	GTX NP or GTX NW	Reduces risk of pumping contamination of the ballast by softened formation soils

Notes :
GTX – geotextile of any type of manufacture
GTX W – woven geotextile
GTX NW – non woven geotextile
GTX NP – non woven needle punched geotextile

Illustrations of some of the applications described in Table 1 follow.

Figure 1, Separation GTX Hong Kong airport

Figure 2, Copenhagen metro over-steep reinforced embankment

Figure 3, A2/M2 J1 to 3 -reinforced slope using geogrid

Figure 4, Soil nailing with erosion control matting

Figure 5, Laying GCL barrier in drainage run-off ponds

Figure 6, Slip of cover soil on steep barrier covered slope

Figure 7, Pond with soil covered barrier & protection GTX

Figure 8, Laying a GTX floating cover over a silt pond

Figure 13, Reinforced soil by geogrid with pre-cast concrete facing

Figure 14, Perforated drain wrapped in a geo-composite drain

Figure 7, Pond with soil covered barrier & protection GTX

Figure 8, Laying a GTX floating cover over a silt pond

Figure 9, Multi-layer barrier/ protection/ reinforcement/ gas venting system

Figure 10, Laying ultra lightweight fill (EPS foam blocks)

Figure 11, Geogrid basal reinforcement at base of embankment

Figure 12, Construction of geocell basal mattress

Figure 13, Reinforced soil by geogrid with pre-cast concrete facing

Figure 14, Perforated drain wrapped in a geo-composite drain

Figure 15, Installing prefabricated geosynthetic band drains

Figure 16, Geotextile erosion control system

Specifications and standards

Specifications

The uses of geosynthetics in transportation projects are either specified in standard national specification documents with project specific annexes or in project specific specifications written by the designers.

For Highway Works the standard specification is the Specification for Highway Works (2) under Clause 609 and Appendix 6/5 for Separation and Appendix 6/9 for Reinforcement and other applications. Additionally requirements for designing reinforcements are given in HA68/94 (3) and for widening highways in HA43/91 (4). Where geosynthetics are used in drainage applications in highways details are included in HA 39/98 (5), and HD 33/96 (6).

For Railway Works the Network Rail Line Standards RT/CE/C/008 General Specification Clauses (7), RT/CE/C/006 Lineside Drainage Systems (8) and RT/CE/S/010 Geotextiles (9) include requirements for the use of geosynthetics as separation and drainage layers.

In addition to the specification documents BSI have published a number of testing standards for Geosynthetics. The initial work was the publication of BS 6906 in 8 parts (10), all of these standards have now been, or will be in the next few months, superseded by equivalent European Standards (BSENs).

The suite of European Standards includes 15 Application Standards (11) (ten for geotextiles and related products and five for barrier products) which set out the requirements to be followed by manufacturers for certification and CE marking of their products in accordance with the Construction Products Directive (CPD). The application standards are supported by a large number of standards (currently 43 are either full ENs or ENV's) which describe test procedures for the determination of product characteristics which can be used in design, identification or specification of geosynthetics. At the time of writing the requirement to supply and use only geosynthetics which are CE marked is not mandatory in the UK, Ireland or Portugal, however, in other EC countries it is a mandatory requirement supported by national laws or regulations. When the UK government enact the regulations the use of CE marked geosynthetics will be enforced by the DTI and Local Trading Standards Officers.

Standards for design

The only British Standard which includes details of design methods for geosynthetics is BS8006 'Reinforced and Strengthened Soils' (12). BS 8006 includes design methods for applications including :

- Reinforced hard faced walls
- Over steep soil fill slopes with soft facings
- Basal reinforcement for fills over soft ground
- Reinforcement over cavities and mine workings
- Reinforcements for Load Transfer Platforms over piles and load support columns in soft ground.

The Design Manual for Roads and Bridges (DRMB) includes some guidance for the selection of filters and separation geotextiles in HD33/96 Surface and Sub-surface Drainage Systems for Highways (6).

Beyond these documents designers will need to consult standard text books, such as Designing with Geosynthetics by R M Koerner (13), manufacturers design guides, (Don & Low, Terram and Tensar have all produced good booklets), or conference papers from the various International Geosynthetics Society (IGS) international and regional conferences. There have been seven international conferences, two European conferences, two Asian conferences and a large number of North American Conferences, all of which have generated a large number of papers describing applications, research, design methods and other topics.

Conclusions

Using geosynthetics in Transportation projects has become a relatively common place feature of the works and there are very few projects where a geosynthetic product is not used in either a temporary or permanent part of the works. Some of the benefits to projects can be summarised as:

- Separation of earthworks materials allows construction to proceed with minimal mixing of soils
- Filters allow the import of drainage media to be minimised
- Reinforcements allow the use of over-steep slopes to reduce land takes and also the use of poorer quality fills.
- Basal reinforcements enhance stability on poor ground, provide an insurance against wholesale failures into cavities and mine workings
- Barriers allow new works to be separated from contaminated ground, avoiding the need to remove the contamination.
- Barriers allow potentially contaminated surface water run off to be kept separate from ground water until it can be treated for a safe discharge.

The overall view is that geosynthetics allow designers to find solutions to problems which are cost effective, can in some instances be considered more sustainable than the alternatives they replace and often allow aesthetic solutions to be engineered to fit into the landscape.

References

BS 6906 Parts 1 to 8, British Standards, London.

BS8006, Strengthened and Reinforced Soils, British Standards, London 1995

BS EN 13249 to 13257 and 13265, Application Standards for the Characteristics of Geotextiles and Related Products, British Standards, London, October 2001

HA 43/91 Geotechnical Considerations and Techniques for Widening Highways, DRMB, Highways Agency 1991

HA 68/94 Design Methods for Reinforcement of Highway Slopes by Reinforced Soil and Soil Nailing Techniques, DRMB, Highways Agency, 1994

HA 33/96, Surface and sub-surface Drainage Systems for Highways, Highways Agency 1996.

HA 39/98, Edge of Pavement Details, Highways Agency 1998

Koerner R M, Designing with Geosynthetics 3rd ed, Prentice Hall, New Jersey, 1994.

prEN ISO 10318 Terms and Definitions for Geosynthetics, British Standards London 2002

RT/CE/C/006, Design Installation and Maintenances of Lineside Drainage Systems, Railtrack (now Network Rail), October 2002.

RT/CE/C/008, General Specification Clauses, Railtrack (now Network Rail)

RT/CE/C/010, Geotextiles, Railtrack (now Network Rail), October 1996.

Specification for Highway Works, HMSO, London 1998

Geotechnical aspects of large section jacked box tunnels

Dr. D. Allenby
Edmund Nuttall Limited
J.W.T. Ropkins
John Ropkins Limited

Introduction

Jacked box tunnelling is a non-intrusive method of constructing a new underbridge, culvert or subway beneath existing infrastructure, for example roads, railway tracks and platforms.

The paper has two main objectives.

- Firstly, to explain to the reader the jacked box tunnelling method, including the techniques used to control ground disturbance, the assessment process required in evaluating a scheme and the prediction of ground behaviour in the design process.
- Secondly, using a series of actual contracts the reader is shown how a variety of geotechnical conditions have been encountered and successfully accommodated in the practical implementation of this highly effective construction method.

It will become apparent that an experienced jacked box tunnel engineer uses his broad based and highly specialised "underpinning knowledge" of geology, soil mechanics, structures, tunnelling, etc to analyse and predict the three-dimensional behaviour of ground and to design accordingly.

The jacked box tunnelling method

Development

The method, as developed in the UK, has its origins in the pipe jacking technology of the late 1960s, when a number of rectangular section pedestrian tunnels and bridge abutments were installed using pipe jacking equipment and techniques. These early examples used precast concrete sections and were installed at shallow depth through highly variable and weak ground conditions.

Transportation geotechnics. Thomas Telford, London, 2003

When a box section is jacked through the ground at shallow depth it does not benefit from soil arching support effects associated with pipe jacking, instead it carries the full overburden and superimposed loads on its flat roof and transmits them into the ground below. Ground closure occurs along the sidewalls inducing a horizontal load whose magnitude and distribution are dependent upon the physical and time dependent properties of the ground.

As the box is jacked forward friction develops on the box/ground interfaces and unless controlled will result in the surrounding ground being dragged forward, ultimately causing disruption to adjacent and overlying infrastructure.

A number of anti-drag systems (ADS) were developed to control and minimise drag effects, and in the early 1970s the first large section boxes were installed. These include a road tunnel at Brent Cross, London, measuring 9.54m wide, 6.86m high by 43.47m long, installed as a three section box with a rear jacking station and two intermediate jacking stations.

Further developments led to the highly successful proprietary wire rope anti-drag system, which together with specialist jacking equipment has made it possible to install a new generation of jacked box tunnels, as shown in Table 1. Milestone achievements include the Docklands Light Railway box at Lewisham Station, London, measuring 17.0m wide, 6.2m high by 48m long, jacked as a monolithic box at a severe skew angle under a live commuter station with 1.7m of cover to the rail sleepers. Recently the technology has been used to install three highway boxes in Boston, USA, through challenging ground conditions. The largest box measured 24.09m wide, 10.8m high by 106.8m long.

The process

In its simplest form, a monolithic reinforced concrete box is cast on a jacking base, typically inside a sheet piled jacking pit adjacent to a road or railway embankment, see Figure 1a. The jacking pit is carefully positioned to minimise the jacking length whilst at the same time satisfying working clearance requirements to the road or railway.

A purpose designed tunnel shield is cast on to the leading end of the box, and thrust jacks are provided at the rear end reacting against the jacking base. During jacking the jacking thrust is transmitted into the ground through the shear interaction developed between the jacking base underside/ground interface, and jacking pit wall/ground interface. Placing and compacting tunnel spoil on top of the jacking base can enhance shear interaction at the underside/ground interface.

In both cohesive and granular material face stability is maintained and ground loss controlled using shield tunnelling techniques. Proprietary anti-drag systems (ADS), are installed at the top and bottom of the box to minimise disturbance to the surrounding ground and possible disruption to the infrastructure above, these are discussed in the following section.

The box is jacked to the headwall and using a carefully controlled and phased sequence the headwall loading is progressively transferred to the shield structure as the box advances and the headwall is removed in sections.

Table 1, Jacked box tunnel projects

Projects	Size	Cover	Date	Ground Conditions	Ground Treatment
Pedestrian and cyclist subway Didcot, Oxfordshire, UK	30m long 5.9m wide 3.6m high	1.7m	1989	Silt-stone fill overlying soft clay	None
Highway tunnels, West Thurrock, Essex, UK	30m long 16.5m wide 9.5m high	8.0m	1991	Chalk with swallow holes loosely filled with sand	None
Highway underbridge, Silver Street Station, London, UK	Twin tunnels each 44m long 12.5m wide 10.5m high	7.0m	1995	Water-bearing gravels above over-consolidated clay containing sand layer with water under artesian pressure	Grouting of water bearing gravels. Dewatering of sand layer
Rail tunnel, Lewisham Railway Station, London, UK	48.0m long 17.0m wide 6.2m high	1.7m	1998	Loose silt and sand overlying soft clay	None
3No. subways, Lewisham Railway Station, London, UK	Up to 32.0m long 4.4m wide 3.65m high	2.0m	1998	Loose silt and sand overlying soft clay	None
Flood relief culvert Dorney, Berkshire, UK	50.0m long 23.0m wide 9.5m high	6.0m	1999	Clayey granular fill overlying water bearing sands and gravels, overlying weathered chalk.	Ground freezing
3No. highway tunnels, Boston, Massachusetts USA	Up to 106.8m long 24.0m wide 10.8m high	6.0m	2001	Weak water bearing strata with numerous man made structures.	Ground freezing
Highway underbridge, M1 Junction 15A, Northamptonshire, UK	45.0m long 14.0m wide 8.5m high	1.6m	2002	Pulverised fuel ash and clay fill overlying stiff clay with rock inclusions	None

Tunnelling then commences by carefully excavating 150mm off the face and jacking the box forward a corresponding amount, see Figure 1b.

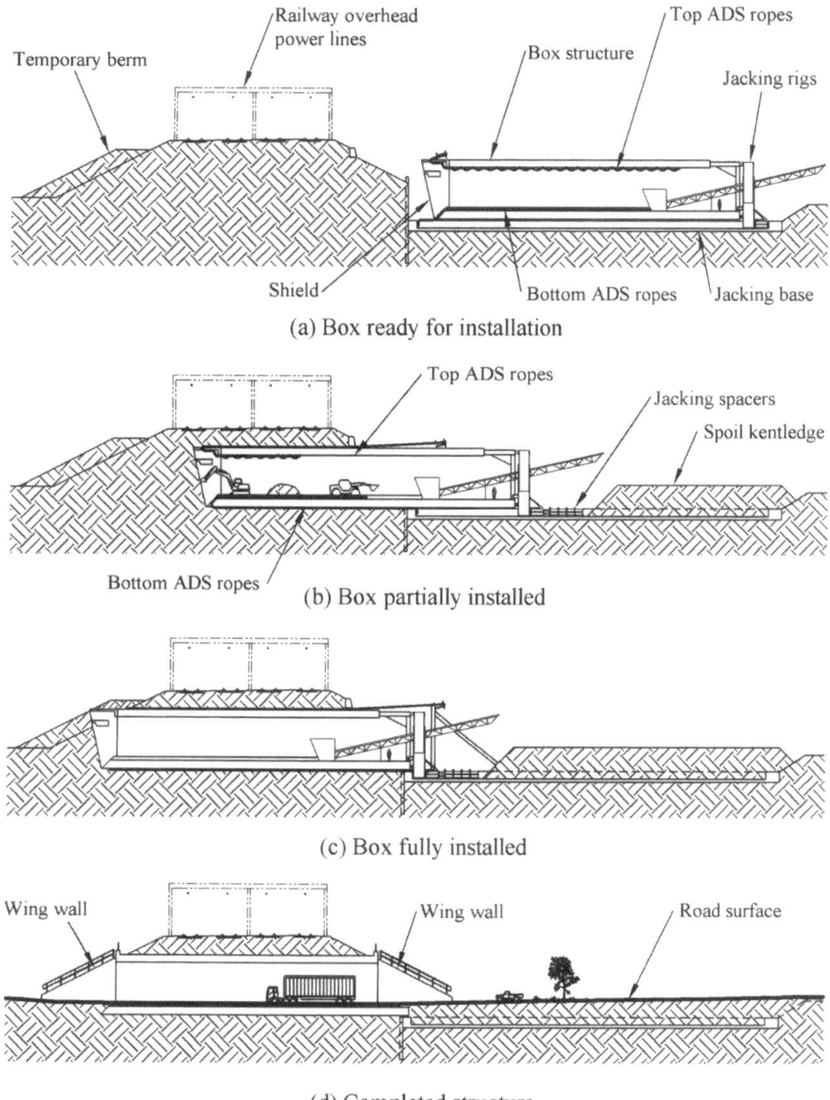

Figure 1, Jacked box tunnel installation

An earth berm is constructed on the exit side of the embankment to buttress the embankment and prevent distress during the final stages of jacking. When the box has reached its final position, see Figure 1c, the box/ground interface is grouted to minimise settlement, the shield, jacking equipment, etc are dismantled, and the portal structures, internal finishings and roadway are constructed, see Figure 1d.

Control of ground disturbance

Surface settlement is an unavoidable consequence of conventional tunnel excavation. It results in a shallow trough formation developing over the tunnel alignment whose magnitude and extent are dependent upon the physical and time dependent properties of the ground, the effectiveness of face support, the speed of erection of the support system and the effectiveness of its primary grouting, and the rate of tunnel advance.

In jacked box tunnelling the box/ground interface cannot be grouted until box installation has been completed, consequently there is a potential for increased time dependent settlement. This may be further increased by drag or shear effects causing remoulding of the ground with a subsequent loss in ground volume.

In some applications, ground treatment or geotechnical processes have been necessary to control ground water, and/or stabilise unstable ground, see Table 1.

The principal measures used to control and minimise ground disturbance are as follows.

Constructional tolerances

To minimise ground drag, jacking loads and settlement it is essential that all external box surfaces are straight, smooth, free from steps and defects, and that opposite faces are parallel and free from twist and distortion. These requirements are achieved by rigorous control of setting out, the use of high quality shuttering materials, checking finished surfaces and rectification of defects.

It is standard practice to accurately define the three-dimensional shape of the box and shield and make adjustments to the trimming beads attached to the shield cutting edges to allow free passage of the box through the ground.

Clearly oversize trimming beads increase overcut around the box sides, and/or roof, with the potential for increased settlement.

Control of face loss

Face excavation causes a three-dimensional stress redistribution in, ahead of, and around the advancing face accompanied by ground relaxation. The ground relaxation element is controlled by buttressing the face using shield tunnelling techniques.

Figure 2 illustrates a composite reinforced concrete and steel shield for a 17m wide by 6.2m high box designed for a mixed face comprising loose silt and sand in the top half and soft clay in the bottom half. The steel shield has a sloping face and hood section designed to be thrust into the loose silt and sand to provide face and roof support and protection to the miners, while the lower concrete shield with its relatively thick walls is designed to support the soft clay. Each top compartment was hand mined and 360° backacter excavators positioned on the box floor excavated the remaining face area.

Shield design calculations must take into account face loads, overburden and superimposed loads, lateral loads and jacking loads including localised loads from face rams, gun struts, etc.

Figure 2, The composite reinforced concrete and steel shield, Lewisham Station, London

Control of ground drag

Referring to Figures 1b and 1c, it will be seen that as the box is jacked forward there will be a tendency to drag the ground. In the case of a wide box with low cover the ground on top of the box could be dragged forward, causing major disturbance and possible disruption to the overlying infrastructure. Similarly, the underside of the box will tend to drag and shear the ground, resulting in remoulding accompanied by a loss in volume causing the box to dive.

These effects can be minimised by using proprietary top and bottom "anti-drag systems" (ADS) as illustrated in Figure 3 to effectively separate the external surface of the box floor and roof from the adjacent ground during tunnelling. They comprise arrays of closely spaced greased wire ropes anchored to the jacking base with their free ends passed through guide holes in the shield and stored with their free ends inside the box. As the box advances the ropes are progressively drawn out through the guide holes in the shield and form a stationary layer between the moving box and the adjacent ground. The drag forces are absorbed by the ADS and transferred back into the jacking base. In this manner the ground is isolated from the drag forces and remains largely undisturbed.

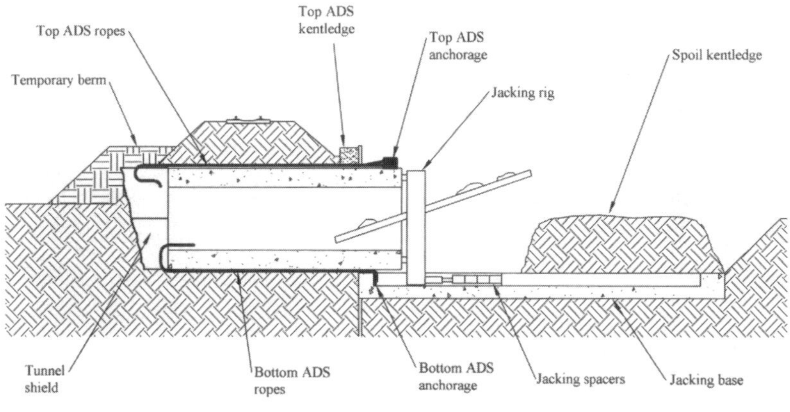

Figure 3, Top and bottom anti-drag systems (ADS)

Lubrication

Friction and drag effects are minimised by adequately greasing the ADS ropes with a good quality grease during their installation, and regularly greasing them during the actual jacking process through arrays of grease injection holes and spreader channels cast into the extrados surfaces of the box.

Ground drag on the sides of the box is normally reduced by lubricating with bentonite slurry.

Project assessment

An experienced jacked box tunnel engineer will visit the site, examine all ground, archive, infrastructure and topographical information available and based on his underpinning knowledge and sound engineering principles, make a judgement as to whether a jacked box tunnel is feasible and the manner in which it should be undertaken.

The assessment process is typically:

- what is the box application, its dimensions and ground cover
- what are the strength and stability characteristics of the ground
- where is the ground water
- site access routes and the most suitable position for the jacking base
- topography and infrastructure, including condition surveys
- how will box installation affect the overlying infrastructure
- can the strength and stability of the ground be improved by adopting a geotechnical process such as dewatering, grouting or freezing, and how will this affect overlying infrastructure
- what face support will be required and how will the face be excavated
- what anti-drag measures will be required and can their forces be dissipated
- anticipated jacking loads and can they be transmitted through the jacking base
- what will be the magnitude of the ground movements and their influences on overlying infrastructure

Clearly, experience prioritises all of the information available, highlights shortfalls and identifies any additional information necessary to develop the project from conception through to completion.

Prediction of ground behaviour in design

The jacked box tunnel designer requires geotechnical input in order to predict the behaviour of the ground and to design both the tunnelling system and the permanent works. This would normally include the specification of a detailed site investigation and its subsequent interpretation.

Interpretation of the site investigation will provide the following information:

- soil types and their elevations
- in-situ densities
- permeabilities
- short-term (undrained) strength parameters for design of the tunnelling system
- long term (drained) strength parameters for design of the permanent works

Additionally, it will determine the elevation(s) of any ground water table(s) and the nature and extent of any buried obstructions.

With regard to the tunnelling system the main aspects of design in which geotechnical input is required are as follows:

- tunnel shield and method of face excavation

- estimate of jacking thrust required to advance the shield and box through the ground
- top ADS
- bottom ADS
- provision of reaction to the jacking thrust from a stable mass of adjacent ground
- soil movements induced by tunnelling

Tunnel shield and method of face excavation

In natural soils a cellular shield configuration is normally adopted featuring internal walls and shelves that divide the face into compartments. The walls and shelves are used to buttress the soils at the tunnel face to ensure their stability, control movements and limit surface settlements. The shield also provides safe access to the tunnel face for miners and machine excavators and egress for the anti-drag systems. A cutting edge at the front of the shield around its perimeter is used to accurately cut the hole through which the body of the shield and the box structure passes.

Soils in the tunnel face must be capable of arching between the internal walls and shelves. Clay soils are normally capable of arching by virtue of their cohesive strength. Most granular soils in the UK are moist and contain some silt and clay particles and as a result exhibit a degree of effective cohesion. If poorly graded granular soils are encountered they may require grout treatment to give them sufficient shear strength to enable them to arch.

Face stability in most natural soils is time dependent, therefore, it is essential to install the box in a continuous tunnelling operation.

A jacked box tunnel is typically advanced in 150mm increments. To achieve the increment of advance the shield must be designed to fail the ground local to the cutting edges, walls and shelves. The failure load must be greater than that required to support the tunnel face, but less than that which could cause heave of the ground ahead of the shield.

If the water table in granular soils lies above the box invert it would normally be lowered to below the box by controlled de-watering techniques. Alternatively, it may be appropriate to stabilise such soils and control water inflow by grout injection in advance of tunnelling.

In some circumstances it may prove economical, or necessary, to freeze the ground in advance of tunnelling in order to achieve a stable face. This is particularly useful when a combination of high water table, weak soils and buried obstructions are present. As frozen soils are typically very strong a vertical free standing face can be worked. The shield is no longer required to support the face but its perimeter cutting edge must be strengthened considerably so as to be able to trim the frozen soil. The increment of advance can be increased to around 600mm.

Geotechnical input is required in the assessment of soil arching capability, the calculation of the face support load and of the embedment load required to advance the shield into the tunnel face.

Calculation of jacking loads

The jacking load is a combination of the shield embedment load and the drag loads at the top, bottom and sides of the shield and box structure.

In frictional soils, and where the box is separated from the soil by anti-drag ropes, the drag is frictional and a function of contact pressure. Where cohesive soils are in contact with the box the drag is a function of adhesion. Both friction and adhesion are normally reduced by lubrication prior to, and during tunnelling.

Geotechnical input is required in the calculation of soil pressures on the box and in the assessment of adhesion.

Design of top ADS

The top ADS ropes are arranged across the full width of the box at close centres, so as to effectively separate the overlying ground from the moving box. The frictional drag induced by the moving box causes a build up of tensile load in the ADS. In calculating the development of this load account is taken of any restraint to movement of the prism of ground overlying the box from adjoining ground.

Geotechnical input is required in calculating available shear and passive restraint on the sides and front of the prism respectively.

Design of bottom ADS

The bottom ADS ropes are typically arranged in two 'tracks' of ropes placed shoulder to shoulder forming slide paths for the box to travel on.

Geotechnical input is required in calculating the allowable bearing pressure on the soil underlying the rope 'tracks' and the settlement likely to occur.

Reaction to jacking thrust

Reaction to the jacking thrust is provided in part by the top and bottom ADS, but mainly by shear interaction between the jacking base and a stable mass of adjacent ground.

Geotechnical input is required in calculating the reaction available.

Ground movements

The ground movements caused by a jacked box tunnel are a function of face loss, over-cut and residual drag, all of which are system specific. Whereas guidance can be obtained from theories of surface movements caused by conventional deep tunnelling the only reliable data for shallow jacked box tunnels is that obtained by back analysis of the surface movements caused in jacked box tunnel projects executed to date.

Our experience on a number of contracts carried out using the methods described above enables us to predict with confidence the development of surface movements during tunnelling and to ascertain whether surface infrastructure can be maintained within acceptable tolerances.

Jacked box tunnel contracts

A number of actual jacked box tunnel contracts are now examined through the eyes of an experienced jacked box tunnel engineer, to illustrate the assessment process, and the development of an appropriate solution leading to a successful box installation.

Didcot, Oxfordshire

A monolithic box 5.9m wide, 3.6m high by 30m long was installed through a four track main line railway embankment to provide a pedestrian/cyclist subway, see Figure 4.

Ground conditions in the embankment comprised 0.6m of track ballast and track formation overlying a loosely compacted siltstone fill founded on natural ground level of soft to firm clay. The ground water table was just below the box invert level.

This was the first application of the proprietary wire rope ADS and of the monolithic box concept which arose from the Client's desire to avoid joints in the subway structure as these would have presented long term maintenance problems.

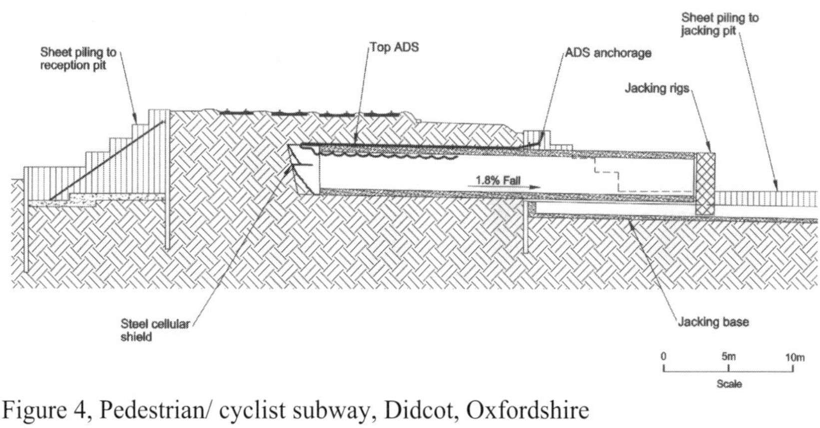

Figure 4, Pedestrian/ cyclist subway, Didcot, Oxfordshire

The box was constructed on a jacking base inside a sheet pile jacking pit with the headwall located in the embankment side slope. The relatively high jacking thrust required to jack the monolithic box was dissipated into the clay ground via adhesion on the underside of the jacking base, and shear/adhesion on the jacking pit side walls.

A purpose designed steel cellular shield with six compartments, three compartments on two levels, was rigidly attached to the leading end of the box. The shield was designed to be thrust into the face to ensure face stability whilst permitting safe working access for miners to carry out the excavation.

The use of a monolithic box resulted in a simple tunnelling operation and ensured an accurate alignment, see Figure 5. Once the jacking pit headwall had been entered the tunnelling operation took just 5 days to complete. Ground movements were so well controlled that it was found necessary to fettle the tracks only twice in order to maintain a temporary line speed of 40mph.

Figure 5, Pedestrian/ cyclist subway, Didcot, Oxfordshire

West Thurrock, Essex

An underbridge 16.5m wide, 9.5m high by 50m long was inserted through a single track railway embankment between two disused chalk pits, see Figure 6.

The embankment comprised loose sand overlying weathered chalk with a proliferation of sand filled swallow holes extending down into the tunnelling horizon. Both chalk pits were dry and records showed that much of the chalk was excavated using bucket equipment operated from the pit floors working high near vertical faces.

As the box was large and located at a cover depth of 8m very high jacking forces were a potential problem. The problem was ameliorated by trimming the sides of the sand embankment thereby reducing the weight of overburden on the box and by using the proprietary wire rope ADS for the first time at the bottom of the box. The reduced overburden weight and the low coefficient of friction on the bottom of the box enabled the total jacking thrust to be limited to 5,000tonnes. The bottom ADS had the additional benefit of preventing crushing of the underlying chalk that would otherwise have occurred with box movement at this highly stressed interface.

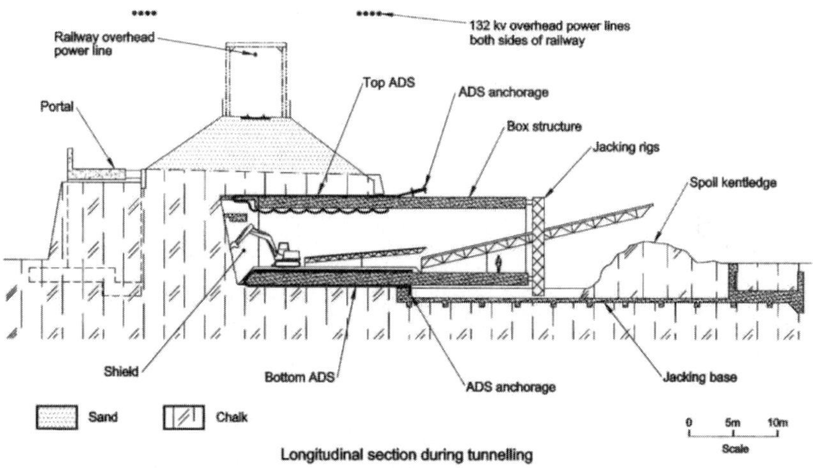

Figure 6, Vehicular underbridge, West Thurrock, Essex

Because the chalk was free standing an open face reinforced concrete shield was designed with eight 2m wide by 2.5m high cells in the top of the shield to permit manual excavation of the sand/chalk interface and to deal with potential loss of sand through the swallow holes. Each cell was equipped with forepoling equipment in the crown, which could quickly be rammed forward into a swallow hole to prevent loss of sand which would otherwise cause settlement, or localised failure of the embankment. The lower face was excavated using 360° backacter excavators operating from the box floor. The shield cutting edge was made robust to cut the weathered chalk without incurring damage.

Anchorage of the top ADS presented a challenge, because there was nothing close by to withstand the 1,000 tonne load. The solution adopted was to tie the anchor beam with steel bar tendons to anchor blocks on each side of the jacking pit which in turn were anchored to the chalk using prestressed ground anchors.

These measures proved to be highly successful enabling the box to be installed safely in 9 days of continuous working with very little surface settlement. See Figure 7.

Figure 7, Vehicular underbridge, West Thurrock, Essex

Silver Street, London

Upgrading of the North Circular Road through the heavily congested Edmonton area of north London required twin tunnels, each 12.5m wide, 10.5m high by 44m long, to be constructed beneath the railway tracks and platforms at Silver Street Railway Station.

This station is a busy commuter station into central London and had to remain fully operational at all times. Jacked box tunnelling was therefore an ideal solution, and the ground conditions and age of the station combined to make this a particularly challenging project.

Limited working space at the station necessitated casting the boxes at ground level as a series of counter-cast interlacing segments, lowering each one in turn into a jacking pit, and incrementally jacking forward, see Figures 8 and 9.

The station structure comprises brick buildings, brick platform walls with arches founded on a made ground embankment. A linking masonry and steel bridge carries the tracks over an adjacent main road and one abutment was within the zone of influence of box insertion.

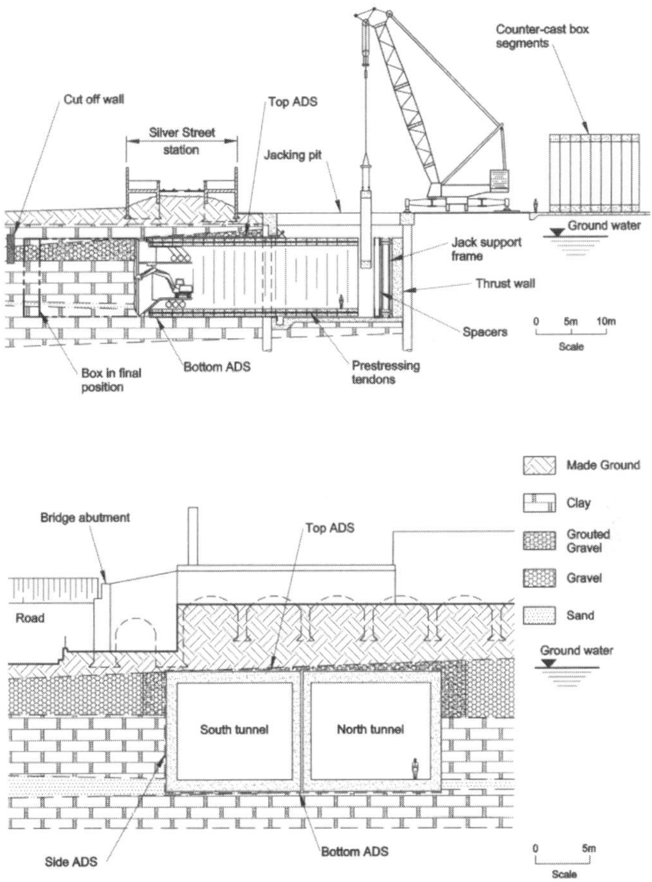

Figure 8, Vehicular tunnels, Silver Street Station, London

Underlying the embankment was a thick continuous bed of waterbearing gravel, overlying London Clay. London Clay is an over consolidated marine clay, regarded by tunnellers as one of the finest materials to tunnel through. At this particular location a 1m thick band of sand under high artesian pressure was present running across the lower part of each box face.

To safely tunnel through these conditions required control of the ground water and stabilisation of the gravel and sand. This was successfully achieved by constructing a jet grouted cut-off wall around the box alignments, extending from the surface down through the gravel bed into the impervious London Clay. The gravel contained within the cut-off wall was pressure grouted using cement-based grouts in a carefully controlled sequence to relieve pore pressure and prevent uplift of the station structures. Horizontal pressure relief drains were installed to relieve the artesian pressure in the sand layer.

Figure 9, Vehicular tunnels, Silver Street Station, London

The proprietary wire rope ADS was used at the top and bottom of each box and for the first time at the side of one box. The second box to be installed was adjacent to the heavily loaded bridge abutment mentioned earlier. To prevent drag induced settlement of the abutment the ADS was used on the exposed side of the second box, which was thrust sideways off the first box during tunnelling so as to maintain full support to the ground on which the abutment rested.

A thrust of 6,000 tonnes was required to install each box and was provided by two vertical thrust walls at the back of the jacking pit. These were structurally independent of the front and sides of the pit and could move back freely when loaded. The design of the thrust walls to accommodate such high loads was critical, as was the assessment of wall movement under load to the design of the jacking equipment.

Both tunnels were successfully installed with settlements being kept within the specified limits set to avoid structural damage to the station. Each tunnel took four weeks to install once the jacking pit headwall had been breached.

Lewisham Station, London

Lewisham Station is built on embankments of very loose, weak silt and sand placed on the original topsoil of the area, which in turn overlies soft clay.

Three boxes were installed two of which were pedestrian subways. The third, probably the most interesting, was a twin cell box carrying an extension to the Docklands Light Railway underneath existing platforms and tracks at a severe skew angle, see Figure 10. The Docklands Light Railway box measured 17m wide, 6.2m high and 48m long with 1.7m of cover to the rail sleepers, and passed under the inner platform foundations and through the outer platform wall foundations. Historically, movements had taken place to the embankment side slopes and platform structure on the jacking pit side.

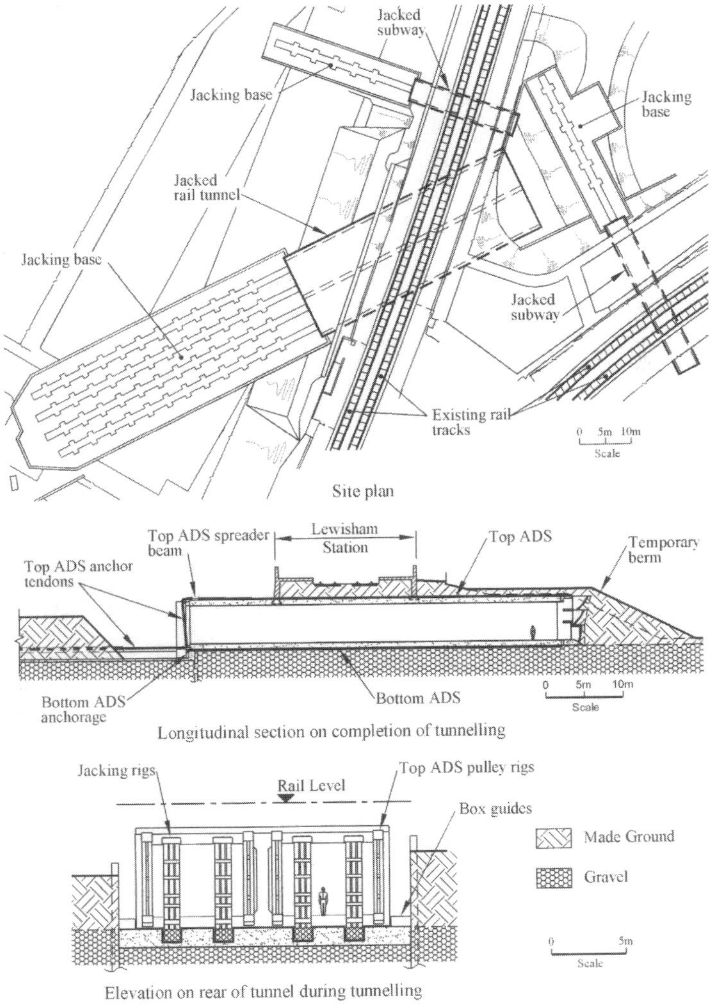

Figure 10, Rail tunnel, Lewisham Station, London

The pre-contract site investigation was supplemented with a series of narrow trenches machine excavated into the side of the embankment, adjacent to the box alignment, to investigate soil conditions and stability within the tunnel horizon, particularly the soft clay. Observations of trench closure showed that the soft clay was stable in the short term, and hand held shear vane tests confirmed its strength.

To successfully tunnel the Docklands Light Railway box through the embankment with 1.7m of cover, required control of a mixed face comprising two distinctly different ground types, together with anti-drag systems to both preserve the integrity of the overlying infrastructure and maintain the box vertical alignment.

A composite cellular shield was designed, see Figure 2. The upper section through the very loose, weak silt and sand comprised a sloping faced steel shield with hand mining compartments on two levels. This sat on a vertical face reinforced concrete shield with steel plated cutting edges and wedge faced dividing walls.

The steel section incorporated the shield used at Didcot, and was designed to penetrate the loose, weak silt and sand inducing face stability through applied pressure and friction developed between the ground and embedded cutting edges, dividing walls and horizontal decks. In addition, the buried cutting edges maintained sidewall and roof stability. The reinforced concrete shield was designed to buttress the face with its dividing walls, maintain sidewall stability and accurately trim the invert to the required profile.

Once the shield had entered through the steel pile headwall the shield's cutting edges, decks and dividing walls, including the wedge faced reinforced concrete walls were maintained embedded a minimum of 150mm at all times. Hand and machine excavation, and jacking followed a carefully controlled sequence with the increment of tunnel advance varying between 75 and 150mm.

In an innovative development the top ADS was anchored via wire rope tendons to the rear of the jacking base. The tendons passed around pulleys mounted in steel frames fixed to the rear of the box. In this way the stability of the box itself was used to transfer the top ADS load of 800 tonnes from a high level down to the jacking base level where it could be readily anchored.

The severe skew gave rise to differential horizontal loads along the box sides and differential embedment loads across the width of the shield. These were accommodated in the design of the side guides and in the planned application of eccentric jacking thrust as required.

Following headwall entry, a three-week period was required to install the box into the position indicated in Figure 11. Ground movements were well controlled enabling the railway and railway station to remain fully operational throughout and damage to the adjacent passenger waiting shelter was avoided.

Figure 11, Rail tunnel alignment and reception area Lewisham Station, London

Dorney, Berkshire

A flood alleviation culvert 23m wide, 9.5m high by 50m long was installed through a 12.0m high railway embankment on the main London to Bristol Railway, see Figure 12.

The natural ground comprised sands and gravels, overlying a chalk aquifer with a water table within 1.5m of ground level. The original embankment was predominantly sands and gravels, but at a later date it was widened on the north side with clay.

The jacking base and culvert box were constructed in a dewatered open jacking pit contained within a perimeter cut-off trench.

To ensure a safe tunnelling process it was decided to stabilise the sands, gravels and clay by ground freezing. An array of horizontal freeze and monitoring holes was directionally drilled through the embankment to freeze a volume of ground extending the length of the proposed culvert and 3m beyond the culvert extrados. Careful grouping and valving of the freeze pipes enabled the freeze growth to be controlled and a frozen cut-off zone to be maintained around the box during tunnelling.

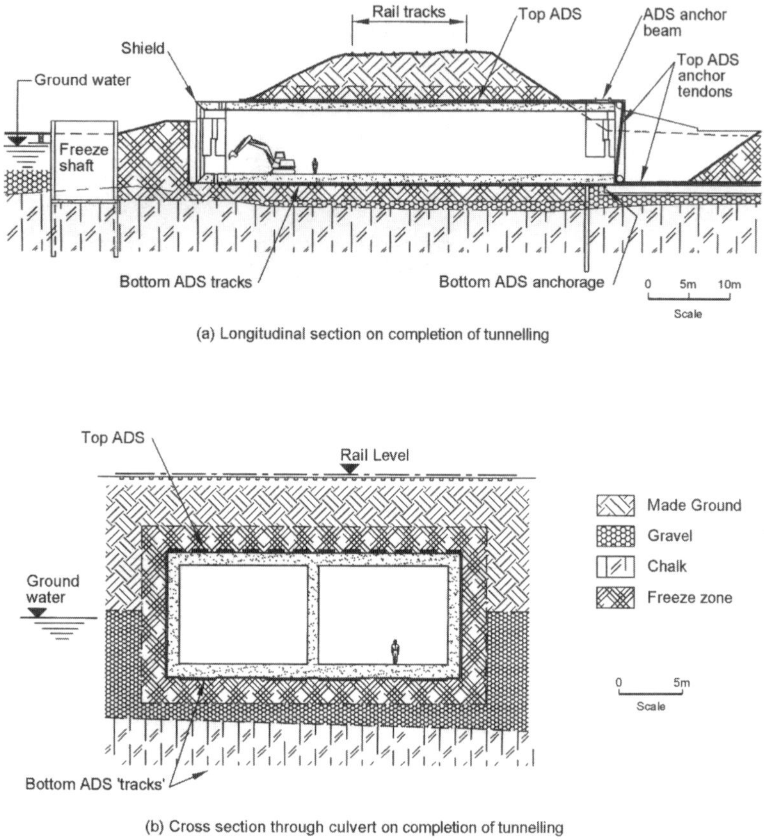

Figure 12, Flood relief culvert Dorney, Berkshire

Figure 13 shows that at the phase change from water to ice there is a dramatic increase in both in-situ compressive strength and Young's Modulus for a number of commonly encountered soil types. Using carefully selected design parameters it was shown that the action of freezing the culvert horizon would create a stable vertical tunnel face without the need for additional face support underneath the railway embankment.

Shield design requirements are significantly different in frozen ground. Although there is no need to support the frozen face it is necessary to ensure that the perimeter cutting edge is strong enough to trim the frozen ground whilst providing safe access to the face for excavation. Figure 14 shows one half of the tunnel face being excavated with roadheader and transverse cutting equipment.

The top ADS was anchored via steel wire tendons to the rear of the jacking base as pioneered at Lewisham. Jacking base stability was achieved entirely by frictional interaction with the underlying sands, gravels and chalk induced by the weight of tunnel spoil placed and compacted on it. Figure 15 shows the box within the dewatered jacking pit.

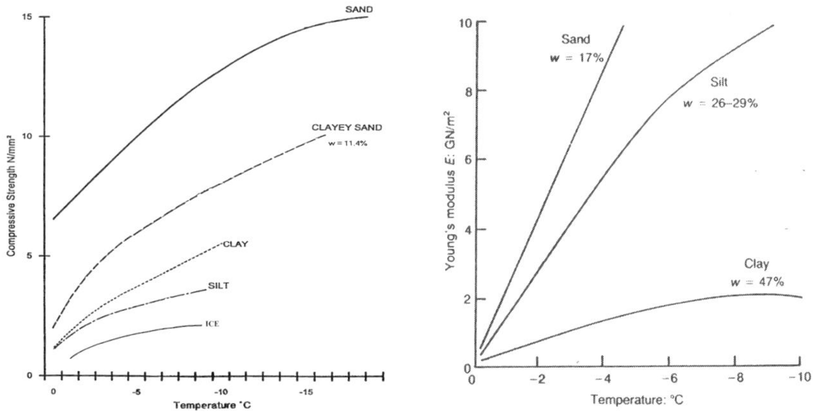

Typical Compressive Strengths of Frozen Soils in Relation to Temperature
(Handbook of Geophysics - US Air Force)

Average Young's Modulus from cyclic loading tests [after Tsytovich (1975)]

Figure 13, Change in soil properties at the phase change from water to ice

A maximum jacking thrust of 6,000tonnes was required to install the box in a tunnelling operation lasting four weeks. Surface settlements were well controlled and permitted rail services to continue uninterrupted at a temporarily reduced line speed of 60mph.

Figure 14, Excavation of frozen sands and gravels, Dorney, Berkshire

Figure 15, Culvert ready for installation, Dorney, Berkshire

Boston, USA

In Boston, USA three interstate highway tunnels were installed under a complex network of seven main line and commuter tracks leading into Boston South Station, to form a vital link in the Boston Central Artery System.

Each tunnel box was installed through the most variable and potentially difficult ground conditions encountered to date without causing disruption to the 40,000 commuters and 400 trains using the tracks each day.

The largest box, I-90 Eastbound, illustrated in Figures 16 and 17, measured 24.09m wide, 10.8m high by 106.8m long.

The site area, shown in Figure 17, had been progressively reclaimed from the harbour over the past 240 years. Across the site lay the buried remains of several old wharves, timber piled structures, stone filled timber cribs, and a depressed trackway comprising mass concrete dressed with granite masonry. Past industry included locomotive repair workshops, an electroplating facility, gas and chemical processing plants, numerous abandoned foundations and a gasholder on the Ramp D tunnel alignment.

Ground conditions comprised a variable fill material over a compressible organic clay with sand and inorganic silt, underlain by Boston Blue Clay. The ground water table lies above the tunnels and is tidally influenced.

In view of the combination of high water table, weak variable ground and the numerous and varied obstructions present, it was decided that ground freezing would be the most effective way of stabilising the ground along each tunnel alignment and providing a ground water cut off. This was achieved using vertical freeze pipes installed from track level.

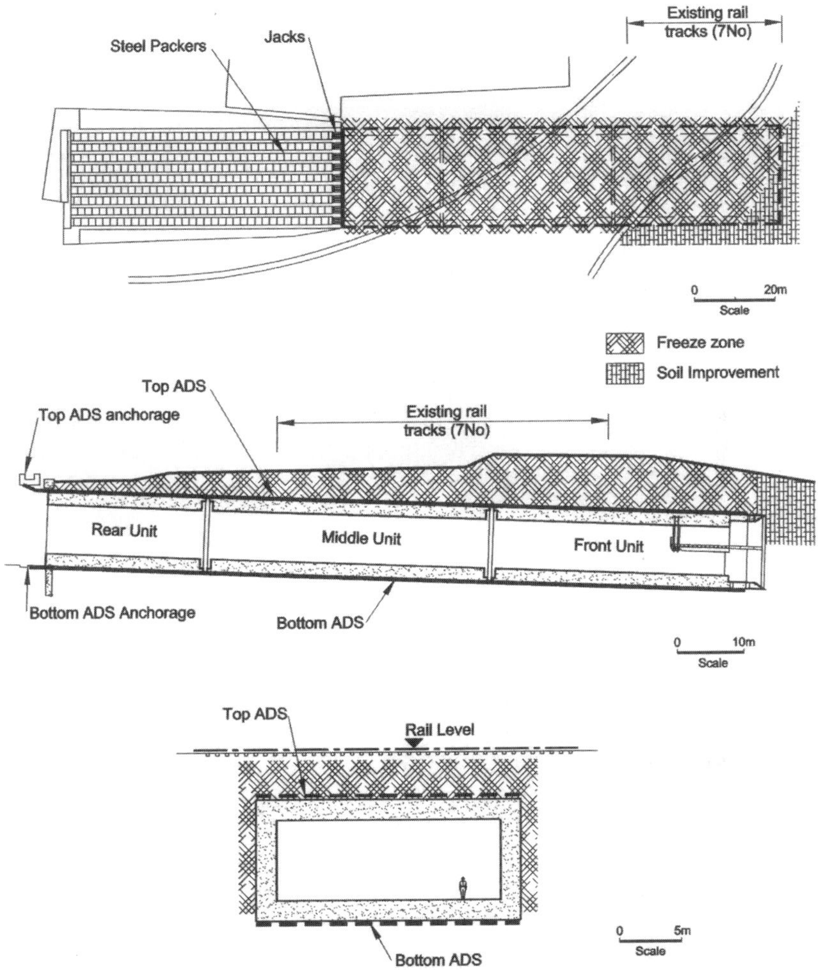

Figure 16, Longitudinal section through I90 Eastbound, Boston

Each box was equipped with an open face reinforced concrete shield divided into two levels with six compartments on each level. Four roadheaders, supplemented with impact hammer equipment were used to excavate the frozen ground and numerous obstructions in each tunnel face.

Ice growth at the sides of each box during tunnelling was carefully controlled to prevent the development of unacceptably high lateral pressures on the box sides. Electric heating cables were installed in the box walls and roof to prevent the box/soil interface freezing and locking the box into the ground.

Figure 17, Boston site, showing the position of the three jacked box tunnels

Figure 18, Rear view of I-90 Eastbound box during tunnelling, Boston

The proprietary wire rope ADS was used at both top and bottom of each box. Top ADS anchor loads as high as 3,000 tonnes were transferred to diaphragm walls forming the sides of the jacking pits.

In view of the very high jacking thrusts necessary to install these large tunnels and the limited capacity of the ground to accept thrust at the jacking pits it was necessary to divide each tunnel into two, or more box units, and employ intermediate jacking stations, (Figure 18). Each tunnel was successfully installed without any delay or disruption to the railway services.

Junction 15A, M1 Motorway, Northampton

A vehicular underbridge 14m wide, 8.5m high by 45m long was installed under the M1 Motorway at Junction 15A to upgrade the existing A43 intersection. Figure 19 illustrates the underbridge box during installation with a minimum cover of 1.6m to the motorway.

The original motorway embankment comprised an engineered clay fill material overlying the natural boulder clay of the Northamptonshire area. After its opening in 1959 the embankment exhibited excessive settlement of the clay fill over a 20 year period, necessitating major reconstruction during the early 1980s. A 3.5m depth of clay fill was excavated under the full width of each carriageway and replaced with pulverised fuel ash carefully laid and compacted in profiled excavations.

Following contract award the site investigation was supplemented with a number of slit trenches mechanically excavated in the embankment side slopes, to investigate the short and long-term strength and stability of the pulverised fuel ash and clay, and to observe the integrity of the interface.

An open face reinforced concrete cellular shield divided into three working levels, each with seven compartments, was designed to support the face. Substantial steel plated cutting edges were attached to the shield perimeter, horizontal working decks and vertical dividing walls, permitting the face material to be penetrated and buttressed to control face loss which would otherwise have led to excessive surface settlement. Figures 20 and 21 illustrate the tunnelling operation and a front view of the shield on completion of tunnelling respectively.

The underbridge was successfully installed in a period of three weeks following shield entry and surface settlements were so well controlled that the 112,000 vehicles using the motorway each day continued unimpeded. Only the southbound hard shoulder and slow lane required resurfacing and this was carried out at night.

This project is the first application of the technique beneath a live motorway in the UK and represents a milestone in the development of jacked box tunnelling.

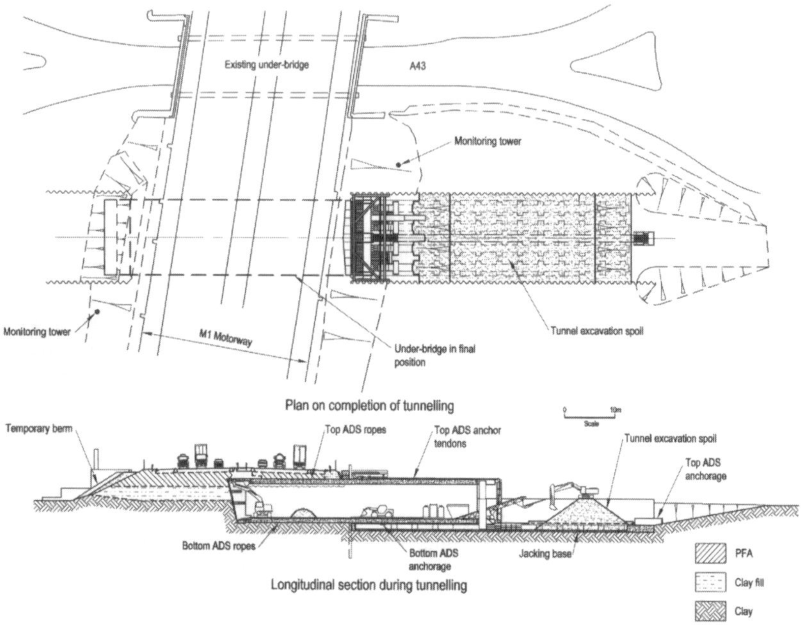

Figure 19, Vehicular underbridge at Junction 15A, M1 Motorway Northampton

Figure 20, Rear view of the vehicular underbridge during installation, Junction 15A, M1 Motorway, Northampton.

Figure 21, Cellular shield and team on completion of tunnelling, Junction 15A, M1 Motorway, Northampton

Summary

The paper has shown that the starting point for a jacked box tunnelling project is a detailed understanding of the ground conditions and geotechnical parameters to be used in the design of the tunnelling system and permanent works. Ideally, these requirements should be achieved through an early involvement and close working relationship between the jacked box tunnelling engineer and a suitably qualified geotechnical engineer.

The greater the confidence the jacked box tunnelling engineer has in the ground conditions and the geotechnical parameters, the more efficient the installation becomes.

A high level of confidence made it possible to install the rail tunnel at Lewisham Station with 1.7m of cover, and the underbridge at Junction 15A of the M1 motorway with 1.6m of minimum cover. Similarly in Boston, USA, confidence in the geotechnical behaviour of frozen ground made it possible to install the longest jacked box tunnels in the world to date.

Based on their experience, the authors believe that the unique and non-invasive technique of jacked box tunnelling will be used widely in the future to install even larger boxes through a greater variety of ground conditions.

Acknowledgements

The authors wish to acknowledge the assistance and encouragement offered by both Edmund Nuttall Limited and John Ropkins Limited in the preparation of this paper.

References

Clarkson T.E. and Ropkins J.W.T. *Pipe jacking applied to large structures.* Institution of Civil Engineers, London, 1977

Ropkins J.W.T. and Allenby D. *Jacked box tunnelling.* Underground Construction 2001, London, 2001

Harris J.S. *Ground freezing in practice.* Thomas Telford, London

Andersland B.O. and Ladanyi B. *An introduction to frozen ground engineering.* Chapman and Hall, London, 1994

Long term settlement of piles under repetitive loading from trains

Dr. N. O'Riordan, A. Ross, and R. Allwright
Rail Link Engineering
A. Le Kouby
SYSTRA

Introduction

The design of piles is an essentially empirical process derived by back-analysis of full-scale static load tests. Results of tests are then related to soil properties, the design proceeds and pile geometries are selected so that the settlements of the piles under design loads are acceptably far from failure. Designers are most comfortable when the majority of the design load is static. This paper describes dynamic load tests on piles and the consequences on design of a piled slab for the Channel Tunnel Rail Link where the dynamic loading from trains was significantly larger than the static or 'dead' load. Ground and pile conditions are related to train loading situations where large permanent settlements could occur. Design methods are described enabling long term settlements of piles to be predicted over the design life of the structure.

Channel Tunnel Rail Link (CTRL)

The Channel Tunnel Rail Link is a new 300 km/hr railway of 104 km length, linking London with the Channel Tunnel. A 7 km length of the railway passes across marshland to the north of the River Thames close to existing ground level at +2 mOD. The ground comprises up to 12m of soft alluvial clays and peats overlying a variably layer of sandy gravels, up to 6m thick. The gravels are underlain by overconsolidated strata comprising, from west to east of the 7km length across the marshes, London Clay, Lambeth Group clays and sands and Chalk. The water table in the alluvium is at approximately +0.5mOD.

Trackbed support on soft ground

The provision of a high speed line across soft ground at a low elevation is fraught with difficulty. Madshus and Kaynia (2001) show how as critical train speed is approached, typically in the range 150 to 200 km/hr, thin or flexible track support systems can undergo very large displacements during the passage of a train. Holm et al (2002) report on the remedial works at Ledsgard, Sweden where shortly after completion in 1997 a speed restriction of 130 km/hr was applied to a length of 200 km/hr line built on a 3 to 4 m thick widened embankment on soft ground.

In order to provide a reliably stable support to the ballasted track for CTRL, a piled slab was devised. After a series of optimisation studies a slab 0.45m thick was selected, to be supported on groups of 4 No piles, 600mm diameter, spaced at 5m centres under the slab. Typical geometries of the slab are given in Montens et al (2003).

The proposed timetable for the operational railway of CTRL produces the loading statistics for the slab as detailed in Tables 1 and 2:

Table 1, Design loading for fatigue analyses

Train	**Axle load kN**	**Mean line load kN/m**	**Minimum axle spacing m**	**Maximum speed km/hr**
Eurostar	170	21.5	3.0	300
Commuter	130	22.5	2.5	160
Heavy Freight	225	80	1.8	150

Table 2, Frequency of loading for fatigue analyses

Train	**cycles/ day**	**cycles/ year**	**weeks/ 100,000 cycles**	**months/ 1million cycles**	**years/ 10million cycles**
Eurostar	1400	511000	10	24	20
Commuter	660	240900	22	51	42
Heavy Freight	750	273750	19	44	37

Derivation of pile loads and results of cyclic pile load tests

In accordance with UIC 776-1R (1994) a dynamic force, (2Qc), of 560 kN per pile was established for Heavy Freight train loading. This can be compared with a static load of around 250 to 400 kN per pile due largely to the self-weight of slab, ballast and ancilliaries. As the ratio of live (train) load to dead (static) load rises above about 0.5, so there is the potential for the development of large long term settlements, see for example O'Riordan (1991). In order to ensure that horizontal and vertical movements were adequately catered for in the design, a unique series of dynamic load tests on selected types of pile was carried out for CTRL.

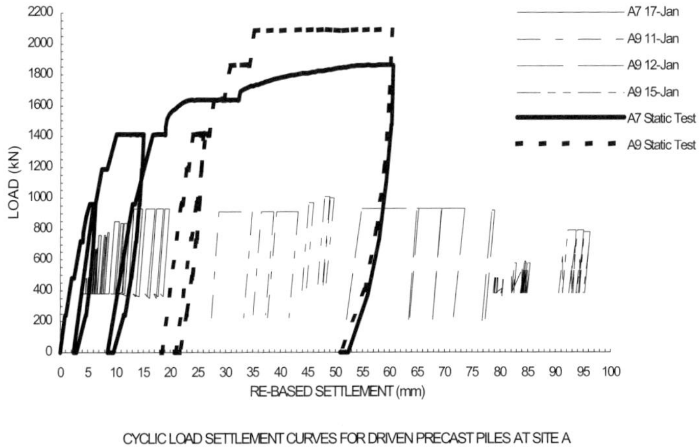

Figure 1, Summary of static and dynamic load tests on 2 DPC piles

The results of static and cyclic vertical load tests on 600mm driven precast concrete piles founded on the gravel stratum are given in Figure 1. Cyclic loads were applied to the pile head at 2Hz. The thicker lines describe the static load behaviour of two 600mm square driven precast concrete piles founded in gravels. For the range of loads predicted in the first instance using UIC 716-1R, it became clear that large permanent settlements would occur, and these initial tests were terminated after a cumulative settlement of about 95 mm was reached. It can also be seen from Figure 1 that reliance on a lumped factor of safety of 2, for example, on the ratio of ultimate (failure) load to total applied load is not adequate to limit long term movement under cyclic loading.

By varying the amplitude of the dynamic loading during succeeding pile tests on driven precast, bored (cfa) and screw piles, it was established that the permanent settlements of piles could be described using equations of the following form:

$$\Delta_N = \Delta_I.B^{\log N} \qquad (1)$$

where Δ_N is the increase of settlement beyond the settlement from dead loading after N cycles

Δ_I is the increase in settlement for the first application of 2Qc, and can be defined as

$$\Delta_I = A(2Qc/Qult)^2 \qquad (2)$$

A,B are empirical constants established for each pile type and Qult is the static ultimate (failure) load of the pile

For screw piles, paired constants A,B were found to be in the range 85, 1.45 and 40, 1.85. Similar ranges in A,B were found for DPC and bored (cfa) pile types in the same ground conditions. These constants will vary with pile type, geometry and ground conditions

Design methodolgy

To refine the calculation of 2Qc, the cyclic load from trains, 3D dynamic finite element analyses were carried out as descrbed by Montens et al (2003). Figure 2 is a visualisation of the train, track, slab and piles modelled principally in LS-DYNA dynamic analyses.

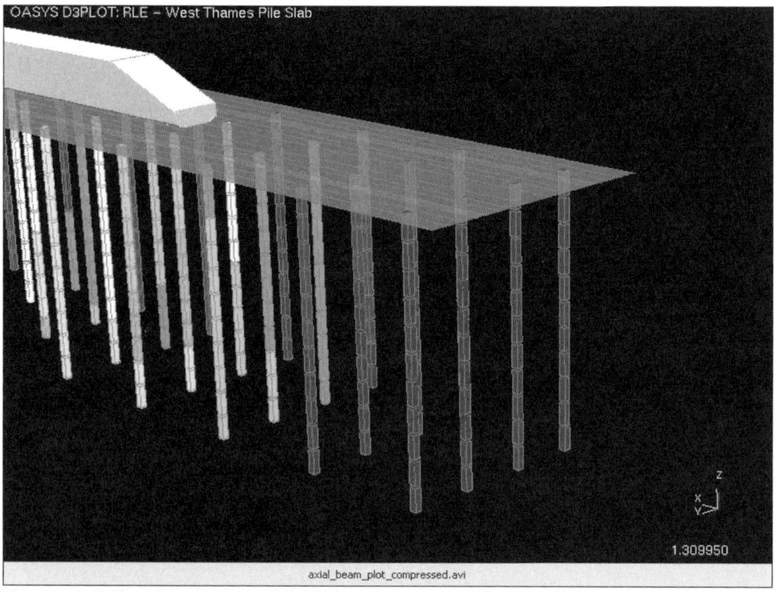

Figure 2, Visualisation of 3D finite element analyses

Shading represents the stresses in the piles during the passage of a train and it can be demonstrated that each pile will undergo slightly different loading depending on its position under the slab. Differences in loading and local changes in ground conditions meant that the calculation of cumulative differential settlements between piles was needed in order to complete the structural design of the slab.

Whilst the track could be re-ballasted to maintain ride quality in the long term, deformations and cracking of the slab due to uncontrolled differential settlements of the piles would tend to go unnoticed during normal maintenance

operations. Consequently a detailed calculation process was undertaken, as follows:

1. select pile type appropriate to ground conditions and site access
2. obtain load spectra from dynamic FE analyses
3. determine Δ_N from eqn.(1) using load spectra for each individual pile under the slab together with the cumulative cycles from Table 1
4. adjust the slab design to accommodate predicted differential movement of individual piles

Figure 3 shows a typical output from LS DYNA for the pile force associated with the passage of a single Heavy Freight train and Figure 4 provides a spectral analysis obtained from this output. Precautionary assumptions were made such that trains passed at identical speeds and their load effects are assumed to be synchronous such that the calculated forces on piles are additive. Long term settlements were found to be dominated by the load effect of passing Heavy Freight trains.

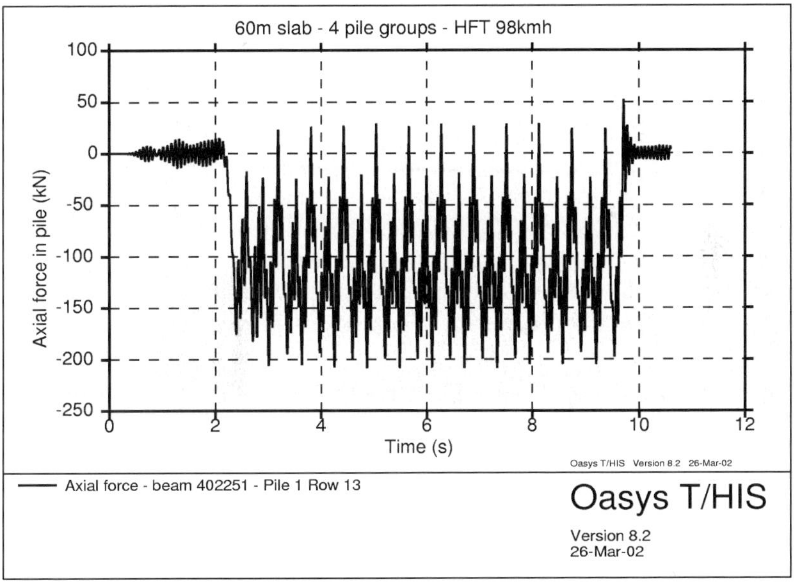

Figure 3, Time history output from dynamic analysis

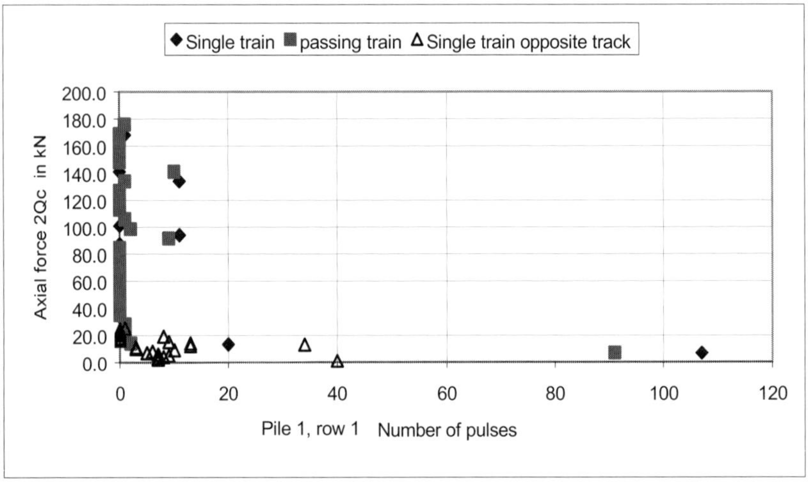

Figure 4, Count of pulses from time history output for single and passing heavy freight trains

From the combination of load spectra from single and passing trains, and using the projected timetable, the permanent settlements were predicted for piles of a given ultimate load capacity. Figure 5 shows the variation of predicted settlement of, in this case driven piles at a particular location under the slab. Similar curves were developed for other pile types. From these fatigue settlement calculations, pile capacities were derived such that the maximum differential movement between piles did not exceed values of between 3 and 5 mm. It was recognised that there would be minor variations in soil properties from pile to pile, and to accommodate this, no allowance was for slab/pile interaction during the development of progressive differential movement: such beam action would tend to reduce differential movements.

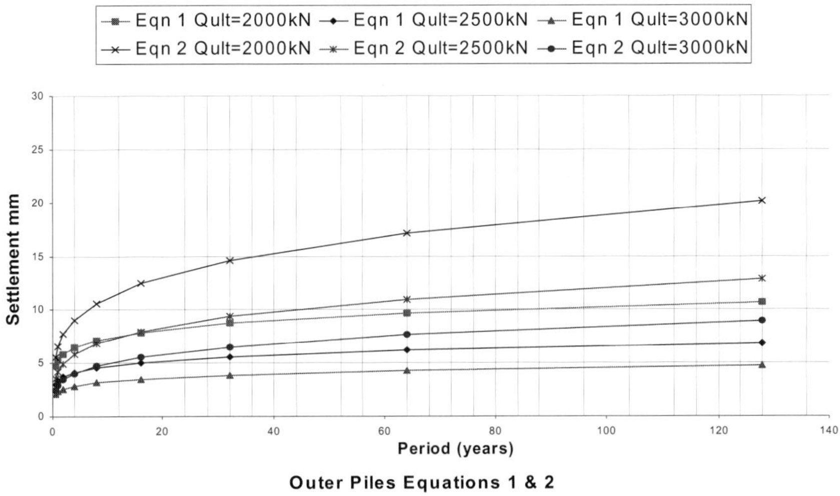

Figure 5, Cumulative settlement of piles of varying ultimate capacity (Note: 'eqn 1' and 'eqn2' refer to extreme values of paired curve fitting constants A, B)

Conclusions

In order to provide a reliable support to ballasted track for high speed railway lines under the onerous geometrical constraints and soft ground conditions covering 7km of CTRL crossing the North Thames marshes, careful analysis of pile loads and associated cumulative settlements was necessary. A design methodology involving full scale, cyclic pile load test results coupled with full 3D dynamic finite element analyses has enabled a structurally efficient and cost effective foundation solution to be found and implemented on site.

It is recommended that, for high speed lines which cross such soft ground conditions, special methods of analysis and design following those presented herein are required where the ratio of live (train) loading to dead (static) load on the support system exceeds 0.5.

Acknowledgements

The work described in the paper has been carried out for Union Railways (North) Ltd by Rail Link Engineering, a consortium comprising Arup, Bechtel, Halcrow and Systra, designers and project managers for the Channel Tunnel Rail Link. The authors acknowledge the help and encouragement from our client, Union Railways, and colleagues within the RLE consortium in the production of this paper. *This paper is a amended version of a paper presented at the Symposium of Structures for High-Speed Railway Transportation, International Association of Bridge and Structural Engineering, Antwerp, 27 to 29 August 2003.*

References

Madshus C and Kaynia A.M. (2001) *High speed trains on soft ground: track-embankment-soil response.* Chapter 11, Noise and Vibration from High Speed trains, ed V. Krylov. Thomas Telford, London.

Holm G, Andreasson B, Bengtsson P-E, Bodare A, Eriksson H (2002) *Mitigation of ground vibrations by high speed trains at Ledsgard,* Sweden. Swedish Deep Stabilisation Research Centre, Report 10. Swedish Geotechnical Inst, Linkoping

UIC 776-1R (1994) *Loads to be considered in railway bridge design.* International Union of Railways. Paris.

Montens S, Leullier J-J, Benadda K and Jenkins S. (2003) *Design of piled slabs for the Channel Tunnel Rail Link.* Symposium on Structures for High-Speed Railway Transportation. IABSE Antwerp (in press)

O'Riordan N.J.(1991) *Effects of cyclic loading on the long term settlements of structures.* Chapter 9, Cyclic Loading of Soils, ed M P O'Reillly & S.F. Brown. Blackie, London

The importance of shallow geotechnics in the performance of rail trackbeds

R.J. Armitage and Dr. M.J. Brough
Scott Wilson Pavement Engineering

Introduction

Traditionally, the Permanent Way Engineer has been concerned with the maintenance of the railway track, essentially those components that could be seen from a visual inspection, walking the track, (although modern technology now means that condition data can be collected from train mounted videos). Similarly, the Geotechnical Engineer would be concerned with the stability of embankments and cutting slopes, the failure of which could not only disrupt the operation of the railway, but also put lives at risk. There was little interaction between these disciplines; indeed the re-organisation of Network Rail still keeps these disciplines apart, the Geotechnical Engineers being considered part of "Structures", which includes bridges and tunnels.

Between the domains covered by these two disciplines lies the Trackbed, which may be defined as the structure (generally comprising ballast underlain by a variety of other materials), that is required to spread the load imposed by trains (via the rails and sleepers), onto the underlying subgrade, whether natural ground or fill. In the last few years it has become recognised that the condition of the trackbed, and the subgrade immediately underlying the trackbed, is critical to the performance of the railway, in terms of the need to maintain Track Quality to an acceptable standard (Hunt, 2000). The domain of the Trackbed Engineer may therefore be considered as covering the top 1-2m of materials, where an understanding of "shallow geotechnics" is crucial.

Such knowledge can only be obtained by studying the performance of actual trackbeds and their constituent materials. During the last six years Scott Wilson Pavement Engineering, (SWPE), have developed a range of trackbed investigation techniques customised to give optimum information within the constraints of operating within the railway environment (Sharpe, 1999; Sharpe, 2000; Sharpe et al, 2000; Collop et al, 2001; Armitage et al, 2003), a range of

Transportation geotechnics. Thomas Telford, London, 2003

interpretative methods have been developed, to ensure this geotechnical data is utilised effectively to optimise maintenance and renewals decisions. This approach has now been used on thousands of miles of track, enabling a broad understanding of the generic causes of trackbed deterioration to be gained.

This paper briefly describes the techniques, before considering two recent case studies where a combination of these techniques (known as Total Route Evaluation) have been used to develop recommendations for Route Enhancements, in one case to take more freight traffic, and in the other case to increase line speeds.

Trackbed maintenance and renewal techniques

The most universally used parameter for assessing the performance of the track is Track Quality, generally calculated from measurements taken with the High Speed Track Recording Coach (HSTRC). This measures the geometric characteristics of the line under vehicle load, and the 35m filtered vertical profile of the left and right rails is of most relevance when judging the performance of the trackbed. Poor geometry will result in a rough ride, not only unpleasant for the passenger, but causing increased dynamic loading and more damage to track/trackbed/vehicle components, with consequent reduction in life.

Track Quality bands are laid down (RT/CE/S/104) for various speed ranges; when exceeded, some form of maintenance will be triggered. The most common technique is tamping, but stoneblowing is increasingly used, and can offer longer life between maintenance interventions than tamping. Ballast cleaning may be appropriate where the lower trackbed layers are good, and the ballast is not contaminated by clay slurry.

However, when a section of track no longer responds to maintenance, it is assumed that some form of renewal is required (incorporating combinations of reballasting, sand blankets, and geomembranes). Up to now such decisions have been made on a reactive basis, however using the techniques described herein predictions can now be made to indicate when tamping will become ineffective, enabling a strategy for Maintenance and Renewal to be prepared for a given route. Therefore the type and extent of a renewal can be optimised, including where it is necessary to maintain existing, or install new, drainage. Site investigation techniques used to facilitate this work include:

- Ground Probing Radar (GPR): A non-destructive technique for assessing ballast depth and condition. Often used in conjunction with an Engineering Walkover Survey, either for preliminary assessment along a route or detailed work at problem sites.

- Automatic Ballast Sampling (ABS): A cost effective sampling technique able to recover all trackbed layers, including ballast and subgrade, even below the water table. Ideal for optimising renewal proposals.

- Laboratory Assessment: Using the samples from the ABS, gradings of granular materials and shear strengths of cohesive layers are commonly performed. In addition chemical analysis enables waste category and options for disposal to be determined.

- Falling Weight Deflectometer (FWD): A method for controlled loading of the trackbed, which enables measurement of both load spreading capacity and critical velocity. Ideal for determining inherent track quality, and hence predicting future track performance.

Ground probing radar (GPR) survey

GPR is used to give an indication of ballast depth and condition. GPR is a non-intrusive technique, which uses electromagnetic radiation to identify the presence of layer interfaces between the different materials comprising the trackbed construction. The equipment records the strength of reflected radiation as well as travel time for the waves. The travel time varies according to the dielectric constant of the trackbed materials, with most influence being due to variations in moisture content. A number of systems are commercially available, with some more effective at assessing trackbed than others.

Where there is a clear interface between clean granular layers (e.g. ballast with air voids) and the underlying layers (e.g. old trackbed or subgrade with a high moisture content), there is normally a strong reflection from the base of the ballast layer. If the ballast has deteriorated or become contaminated, the radar waves cannot penetrate as easily, so the strength of reflections from the base of the ballast will be low. However, if there is a distinct layer of contamination within the ballast (e.g. clay slurry) there may be a strong reflection from its upper surface.

The GPR data is referenced to spatial position on the track using a measuring wheel recording in metres. The positions of key track features are also recorded, including mileposts, bridges and the locations of wet spots. GPR can also be used to highlight potential earthworks and formation anomalies that may not be apparent from other visual inspections.

The GPR data undergoes an initial processing to remove extraneous noise, and an appropriate gain function is employed to produce a greyscale plot. This is combined with Track Quality data obtained from the HSTRC (in digital format) on a pro forma, each one representing half a mile of track, which is subsequently used by the engineer to perform a targeted Walkover Survey. It is often useful to consider four Track Quality bands (in terms of the Standard Deviation, SD, of the profile about a mean).

Table 1, Track quality categories

SD	Track Quality Description
<2mm	Good
2-3mm	Average
3-5mm	Poor
>5mm	Very Poor

Walk over survey

The purpose of the Walkover Survey is to visually examine the whole route, with a view to gaining an interim assessment of the extent of the problems, so that a strategy for further investigation and prediction of the likely requirements for treatment can be proposed. The engineering Walkover records:

- Track form and any defects/discontinuities likely to cause faults in vertical alignment.
- Ballast volume dimensions including shoulder width, shoulder height and crest width.
- Nearby structures (platforms, retaining walls, overbridges, underbridges).
- The topography of the route (cut, embankment, at grade, earth bunds).
- The type and condition of any drainage present.
- Condition of ballast in cribs and shoulders.
- Signs of embankment instability.
- Cess height relative to rail level.

For the Preliminary Phase of Total Route Evaluation (Sharpe et al, 2000), GPR and Walkover Surveys are performed throughout the route and the information combined with HSTRC Track Quality data on Total Route Evaluation plots. Subsequent interpretation allows determination of generic causes of track quality deterioration and definition of targeted locations for physical sampling (with the Automatic Ballast Sampler, ABS).

Automatic ballast sampling (ABS)

When choosing the spacing of ABS locations some general principles are required. Where the trackbed is uniform over a significant length, as suggested by the GPR trace, it is only necessary to take an ABS every 200 metres, or 10 chains. Where there are rapid variations in trackbed condition a spacing of between 50 metres or 100 metres is recommended, depending on the degree of variation and the lengths affected. An ABS spacing of 20m is often employed at overbridges to assess track-lowering options. The samples are taken close to the cess rail in the ballast crib, where the best indication of trackbed conditions under loading occurs.

In order to identify any hidden services, all excavations are made in accordance with RT/LS/S/011 (Issue 3 February 2002). A visual survey is undertaken to identify indirect signs of buried services such as catchpit covers or lineside Signal & Telegraphy (S&T) location boxes. Once the ABS locations are established, the ground is CAT-scanned using a Network Rail approved scanner prior to any excavations. Where susceptible infrastructure is identified, or where in doubt, the location of the excavation is moved to a suitable place.

The ABS recovers representative cores of ballast and lower trackbed layers in a clear plastic liner, which is sealed and then returned to the laboratory without further disturbance to preserve its lithology. Some of these samples are driven deeper, to a depth of 2.0m below rail level, in order to investigate the condition of the underlying subgrade.

Laboratory work

The ABS samples recovered from site are received at the laboratory where they are cut in half to allow detailed logging and photography (Figure 1). This enables the nature of the subgrade and trackbed layers to be accurately recorded, and the ballast condition to be assessed. Detailed logging of the ABS follows a procedure established by SWPE and now endorsed by Network Rail in draft Railtrack Code of Practice for Formation Treatments (RT/CE/C/039 Draft 1d). The depths of any obvious water table, geotextile and/or impermeable layer encountered, are also recorded.

On completion of the logging, the upper part of any ballast layers is removed. Sub samples of similar condition are combined, and then separated using a riffle box; two thirds being allocated for geotechnical testing and one third for chemical testing.

The geotechnical testing regime comprises Particle Size Distribution Analysis (PSD) in accordance with BS1377: Part 2 1990 and Wet Attrition Testing (WAV) in accordance with Railtrack Line Specification RT/CE/S/006 Issue 3 August 2000. This factual data is used to help categorise the ballast in terms of Residual Ballast Life and Ballast Cleaning Potential.

Trackbed layers are then presented on longitudinal sections using colour coded vertical bars, referenced from the top of rail, in order to assist engineering interpretation of the data, as shown in Figure 2. After logging of the ABS samples, and materials testing, a summary of the data is added to the Total Route Evaluation plots, in order to aid engineering interpretation.

Figure 1, Example of ABS photographic log

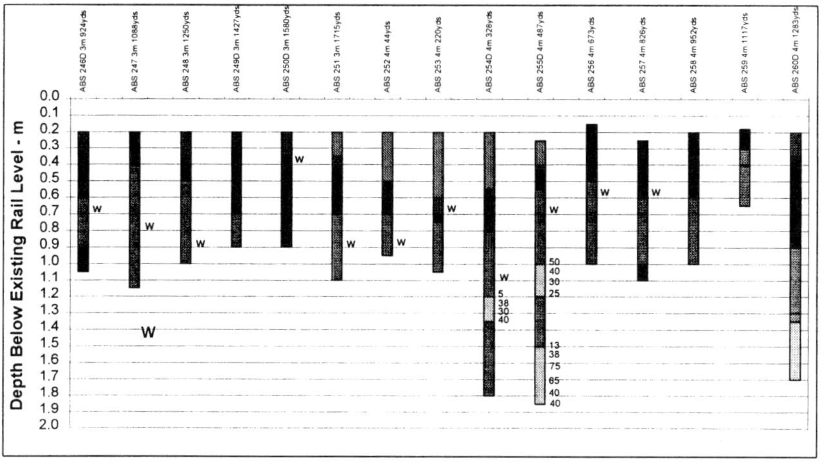

Figure 2, Example of ABS longitudinal section

Falling Weight Deflectometer (FWD)

Where poor stiffness or critical velocity is deemed to be the root cause of track quality deterioration, FWD testing may also be required to further characterise the trackbed. These parameters are discussed later in this paper with specific reference to the case study on the West Coast Main Line.

The FWD is a non-destructive device designed to apply a known impulse load via a loading platen. In its rail operation it applies this load via an unclipped sleeper to the trackbed. Each sleeper is loaded three times with a nominal load of 12.5 Tonnes. The load pulse approximates to that applied by a single axle of a train passing at speed. For each FWD test, the peak load and transient deflections of sleeper and ballast surface are measured to an accuracy of a few microns (mm x 10^{-3}), using geophones.

Following processing of FWD data, and interpretation in conjunction with other trackbed information (from GPR, ABS and Walkover Surveys), decisions can be made as to appropriate maintenance and renewals proposals, line speed restrictions or deeper stabilisation/stiffening methods.

Total route evaluation plots

The Total Route Evaluation plots detail the line profile, including topography, locations of curves, drainage, bridges, crossings, signals and other infrastructure (Figure 3). Standing water and wet beds are also highlighted. Additional information, specifically observations of features that may affect trackbed condition or are a consequence of trackbed deterioration, are recorded (using codes) in tables on the Total Route Evaluation plots. These include:

- Track quality – An engineering assessment to highlight any sections that may have deteriorated due to poor trackbed condition.
- Trackbed faults – Locations of perceived variable support conditions, observed earthwork faults and slip failures, sections exhibiting poor drainage, wet beds and track irregularities. (Track irregularities typically point to anomalies in the rail, track infrastructure such as AWS magnets or breathers, or sections with track components missing or in a poor condition).
- Rail/sleeper type – Locations of rail type (continuously welded or jointed) and sleeper type (concrete, wood and steel). Locations of switches and crossings are also detailed.
- Ballast condition – A visual subjective assessment of ballast contamination (clean to very dirty).
- Drainage – A visual assessment of the type and condition of the cess, drainage channels, catchpit networks and any other relevant features that may affect drainage. Standing water is also highlighted, and any obvious direction of flow of drainage runs.
- Topography – Embankment, cutting or grade.
- Geology – Based on a desk study of relevant geological maps.

With the information combined in a readily accessible format, it is possible to develop recommendations for treatment, as described in the case studies, in conjunction with other constraints that may limit the options for a given route (such as access, possession times or length of blockade, availability of materials and plant.)

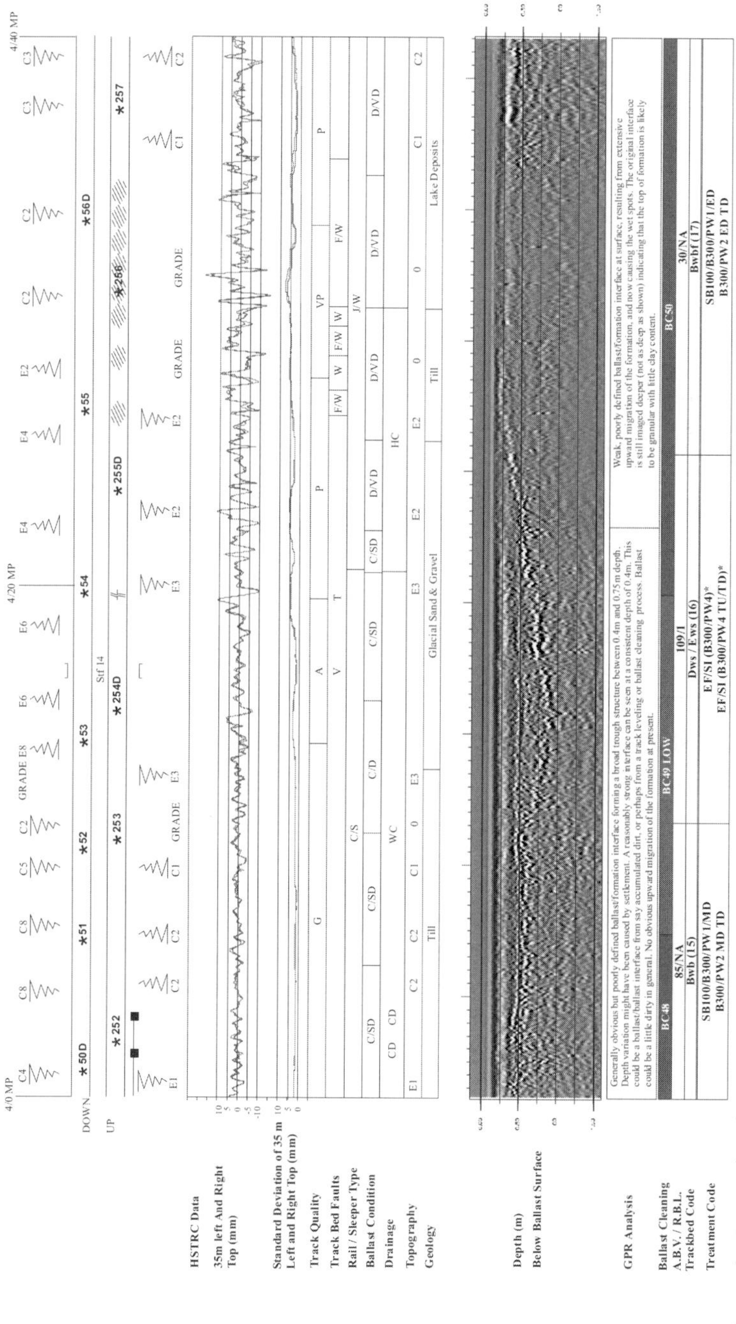

Figure 3: Example of a total route evaluation plot

ECML, Freight primary route 3

FPR3, otherwise known as the Stillington Branch, runs from Norton-on-Tees Junction to Ferryhill South Junction, consisting of 20miles 68chains of relatively little used twin track. The route passes through varied topography, with some cuttings and embankments up to 20m high. The geology is also varied, but drift deposits predominate, comprising Till (Boulder Clay), Glacial Sand and Gravel, Alluvium and Head. The underlying rocks are of the Sherwood Sandstone Group or Magnesian Limestone.

The main objective of the study was to develop an outline scope for upgrading the Stillington branch as a potential diversionary route, to release existing train paths on the East Coast Main Line (ECML) for the use of passenger traffic, or provide an increase in capacity for through freight traffic. It was recognised that some sections would require renewal of trackbed layers (incorporating ballast, geotextiles and sand blanket as appropriate), whilst ballast cleaning might be adequate along some lengths, and regular tamping alone along others. The methodology for making these engineering decisions was based on calculating several key parameters; the first two mostly concerning the ballast condition and the second two the structural adequacy of the trackbed.

- Residual Ballast Life (RBL): an assessment of life in years after which tamping may be deemed ineffective, based upon typical maintenance budgets, existing track quality, ballast grading data (PSD), Wet Attrition Values (WAV), and trackbed condition.

- Potential for Ballast Cleaning (BC): an assessment of ballast condition and depth, with estimated rates of return, based upon ABS and ballast grading data, and typical ballast cleaning operational limitations.

- Trackbed Index (TI): this provides a quick assessment of bearing capacity, using undrained shear strengths from materials testing and trackbed depth. A minimum value of 40 is recommended for a line taking 25T axles (RT/CE/C/039 Draft 1d). If values fall consistently below the required level, it may be necessary to perform Falling Weight Deflectometer testing to determine the actual load spreading ability of the trackbed.

- Average Ballast Volume (ABV): in order to highlight sections of track with excessive ballast volumes, representing potential embankment instability, slip failures and formation settlement. Using measurements obtained during the Walkover, ABV is expressed as a percentage of a standard cross section ($1.7m^2$), based upon typical shoulder height, shoulder width, and ballast depth taken from Railtrack Line Specifications (RT/CE/S/102 Issue 5 Feb 2002).

- Trackbed Code: this summarises the trackbed types and specific characteristics observed on each section of track, focussing on ballast condition, trackbed depth, drainage and evidence of blanketing.

Residual ballast life (RBL)

Residual Ballast Life is calculated using typical maintenance budgets, limiting track quality, existing track quality, ballast grading data and predicted ballast deterioration. The following inputs to the model were assumed:

- Traffic carried between tamps, measured in million gross tonnes (MGT), was based upon the existing maintenance budget (i.e. typically 3 miles of tamping per year was being performed, giving a return cycle of roughly 3 years 4 months) and the annual 4 MGT per annum of traffic.

- Limiting track quality was assumed to be 3.8, based on target standard deviation (SD) values (for eighth mile sections) below which 50% of recorded SDs should fall (based upon Railtrack Line Standard RT/CE/S/104, assuming a maximum line speed of 50mph from the Sectional Appendix data for the Stillington Branch).

- Existing track quality was based upon SDs for eighth mile sections (35m filter).

- Ballast grading data and Wet Attrition Values came from materials testing.

The model calculates an inherent SD at each ABS location, taking ballast condition into account. Algorithms predict ballast breakdown and associated track quality deterioration. The model assumes a tamping intervention is required when the limiting track quality is reached. This process is repeated until the calculated annual tamping frequency is greater than four, at which stage the ballast is classed as life expired, and the RBL is then reported in years.

Residual Ballast Life was only calculated where the trackbed was suitable for tamping. Ballast samples were scrutinised, and if the construction depth of ballast was greater than 150mm and contained no slurried material, the trackbed was deemed suitable.

Potential for ballast cleaning (BC)

In assessing the ballast cleaning potential, several factors were considered, relating to both trackbed condition and ballast cleaner operational limitations. Ballast cleaners should not be considered for use where the ballast contains clay slurry. This material, especially when wet, tends to clog both the cutter bar and the screens of the machine. The depth of influence of the cutter bar was assumed to be approximately 300mm below sleeper bottom. The following criteria were used to assess suitability for ballast cleaning.

- Ballast should extend to a minimum depth of 300mm below sleeper bottom.
- Water should not be encountered within 300mm below sleeper bottom.
- Slurried material should not be encountered within 300mm below sleeper bottom.

As the cutter bar on ballast cleaners can be set at a variable depth, the ABS samples that marginally failed to meet the above criteria were reassessed based on the ABS photographs.

- Where coarse ash (only) was present within the ballast layer but no higher than 250mm below sleeper bottom.
- Where the water table, or some minor slurried material was present, but no higher than 250mm below sleeper bottom.

The percentage of anticipated ballast cleaning returns can be extrapolated from the results of the Particle Size Distribution analysis (in accordance with typical sieve sizes on a ballast cleaner), corrected for material lost during sampling (ie. oversize ballast dislodged due to the internal diameter of the ABS tube) using measurements taken during the site sampling. The results were reported according to one of the following bands.

Table 2, Bands for reporting ballast cleaning rate of return

Ballast Cleaning Rate of Return	Lower limit	Upper limit
High	75%	75%+
Moderate	60%	74%
Average	40%	59%
Low	0%	39%

Condition summary

All trackbed condition data for both the Down and Up lines was summarised in a detailed route commentary, dividing the route into sections with similar trackbed condition. This highlighted potential or existing problems in the trackbed, and possible deterioration mechanisms, which enabled each section to be assigned a Trackbed Code.

Typically throughout the site, trackbed construction comprised ballast in poor condition, contaminated with the breakdown material from ash and clinker, ballast and sleeper degradation. Depth of trackbed was typically sufficient, with up to 26% of track demonstrating good formation with adequate depth of underlying trackbed layers, typically comprising course ash in a fine ash matrix.

Up to 48% of track exhibited a slurried upper trackbed, which could potentially dry out with appropriate maintenance and provision of adequate cess drainage. The remainder of the track (26%), whether with adequate or inadequate trackbed depth, had poor formation and drainage problems. There were potential problems with the bearing capacity throughout 33% of the route.

A major proportion of track exhibited potential earthworks instability, typically identified by excessive ballast depth, step changes in the GPR interface, or observations made of slip sites during the Walkover. These were usually observed in similar locations on the Down and Up line.

Recommendations for treatment (treatment code)

Following the interpretation of trackbed condition, ballast cleaning potential and residual ballast life, decisions were made as to whether to maintain or renew. Where the trackbed was inadequate and tamping was predicted to be ineffective in the medium term, renewal was ideally required, but a reduced specification (i.e. shorter life solution) was developed which included ballast cleaning. In effect the staged trackbed investigation, associated materials testing and subsequent interpretation allowed optimisation of maintenance and renewals to fit the overall budget and route strategy.

Where renewal was required, ballast replacement to give the standard construction depth of 300mm was recommended, with a minimum crossfall of 1 in 40 to the cess or to a suitable drain. Due to the observed slurrying, a 100mm sand blanket was required over 17% of the route. This was typically associated with the need for maintenance or improvement of the existing cess drainage, and extension of drainage.

A range of geotextile separators was proposed for use throughout the route. Not only filter/separators (such as Terram PW1), but also drainage enhancing (eg PW2), and grid reinforced (e.g. PW4). The drainage enhancing filter/separator provides extra abrasion resistance when placed over old granular or ash trackbed layers, as well as a path to remove water filtering through the ballast, so long as the geotextile is connected to adequate lineside drainage. The reinforcing filter/separator can be used to stiffen the ballast structure over weak ground, where increasing the construction depth is not feasible.

Table 3 presents the length of proposed treatments, together with a Treatment Code that was added to the Total Route Evaluation plots.

Table 3, Summary of trackbed treatments

Treatment	Treatment Code	Track Length
Ballast replacement (300mm construction depth)	B300	10m 968yds
Sand blanket (100mm depth)	SB100	1m 1391yds
Separating geotextile (includes filter / drainage enhancing / geogrid)	PW1-4	10m 968yds
Excavate clay formation (to allow 300mm construction depth of clean ballast and a 100mm blanketing layer)	ECF	0m 280yds
Maintain / improve existing cess drainage	MD	4m 448yds
Extend existing cess drainage	ED	4m 976yds
Raise the track	RT	0m 645yds

Although clay materials were encountered in the subgrade throughout the site, they were generally at sufficient depth not to result in poor bearing capacity. Only 1% of the route required the underlying clay to be excavated. In another section with an underlying formation problem, track raising was potentially feasible.

A number of sites (such as at overbridges) were proposed for track lowering in order to increase clearances. Such lowering can result in a range of ongoing track maintenance problems if careful attention is not made to existing ground conditions. Additional sampling was therefore recommended in order to better identify subgrade properties, from which an appropriate trackbed could be designed.

WCML, route speed enhancement

This route consists of about 22 miles of twin track (Up and Down Main) within which there is a 3.5mile section of additional twin tracks (Up and Down Slow). The varied topography included sections on high embankments, and there had been a history of instability at some of these (characterised by considerable thickness of ballast due to frequent topping up as the embankments had settled). The line speed was generally 80mph, but it was hoped to raise this to 125 mph, in order to facilitate running of high-speed trains on the West Coast Main Line.

The main objective of the study was to use Total Route Evaluation to identify those sections of track with track quality deterioration problems, and those where problems could result from "dynamic interaction" at the higher speeds proposed. Dynamic interaction can occur when the speed of a vehicle travelling over soft ground exceeds approximately 70% of the ground wave velocity ("Critical Velocity"). The result can be excessive deflection of the trackbed, resulting in the need for continued maintenance as the Track Quality rapidly deteriorates. In practice, line speed restrictions have been imposed on those sites where such problems may occur, based on past experience. Testing with the FWD enables the Critical Velocity to be assessed, along with the Sleeper Support Stiffness (as described in RT/CE/C/039 Draft 1d), which helps to assess those sections of track with inadequate bearing capacity.

- Sleeper Support Stiffness: an assessment of the response of the trackbed under a dynamic load comparable to that imposed by a 25T axle. An important measure of overall structural adequacy and also the variability of track support, which influences track quality.

- Critical Velocity: an assessment of ground wave velocity. Particularly important on high-speed lines built over soft subgrades where speeds may be limited due to "dynamic interaction".

Sleeper support stiffness

It has been shown that, for good inherent track quality, it is essential for the track to have both adequate and uniform stiffness (Hunt, 2000; Collop et al, 2001). In other words, higher deflections represent lower stiffness (load spreading ability), and track quality deterioration is worse on trackbed exhibiting variable sleeper support stiffness (i.e. variable FWD deflections).

The FWD provides a convenient, relatively quick, method of applying a measured load to an unclipped sleeper, which then transmits the load to the trackbed. Deflection d0 is the average of two deflections measured on the unclipped sleeper near to the rail seats. Deflection d300 is the average of three deflections measured on the ballast, offset 300mm from the centre of the loaded sleeper; deflections d1000 and d1500 are averages of two deflections measured at 1000mm and 1500mm from the load respectively. The Sleeper Support Stiffness can then be calculated and compared with the values recommended in Table 4 from Draft 1d of RT/CE/C/039. For example a value of 60kN/mm/sleeper end would be required for new track carrying trains at up to 100mph.

Critical velocity

On soft ground the phenomenon of 'Dynamic Interaction' can occur under train loading, characterised by increasing elastic displacement as the train speed increases. A theoretical study by Hunt, (1993) and work by Sharpe, (2000) has indicated that the maximum practical speed at which trains can run before track quality (and maintenance activities) are adversely affected is about 70% of the Critical Velocity.

The velocity of wave propagation for any section of track tested with the FWD can be determined from the recorded time histories. This is done by measuring the time lag between the peak deflections at the sleeper and those of the geophones at offsets of 1000mm and 1500mm from the loaded sleeper.

Track quality deterioration codes

Based on data from the Preliminary Phase of Total Route Evaluation, it was possible to identify the key factors governing Track Quality along the length of Route 7:

Track (T): Track component deterioration or transition (i.e. from one component type to another, for example from continuously welded rail to jointed rail).
Ballast (B): Deterioration of ballast condition as a result of mechanical maintenance, ballast attrition under loading or general contamination from other external sources.
Drainage (D): Poor drainage causing deterioration of ballast quality or general water ponding in the trackbed. (Drainage related ballast contamination problems are more apparent from detailed ABS testing).
Formation (F): Poor formation and drainage particularly when considering formation migration, formation pumping or development of wet spots.
Variable Support (V): Change in trackbed support conditions at the beginning or end of an overbridge, underbridge, viaduct or other structure (including drainage pipes, culverts etc.). This also includes perceived variable support conditions due to underlying geology and trackbed/formation condition.
Variable Support (Earthworks/Formation) (VE): Change in trackbed support conditions specifically due to underlying earthworks and formation anomalies. These typically include sudden increases in ballast depth due to past embankment instability. These can often be interpreted from GPR traces.
Infrastructure (I): An isolated area of poor track such as an AWS magnet, insulated block joint, breather, greaser, clamped rail, dipped welds, a Signal or other S&T (Switches and Telecommunications).
None Apparent (N): No apparent factors governing track quality deterioration. (Usually lengths with good track quality).

A summary showing the percentage of track suffering from these types of deterioration is given in Figure 4.

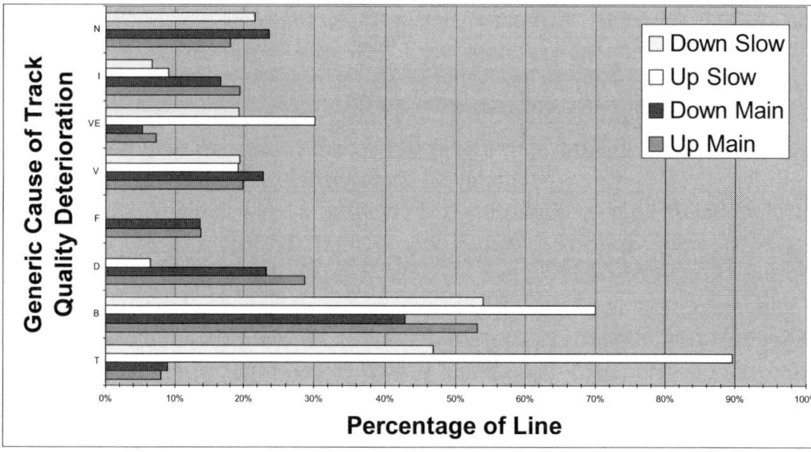

Figure 4, Summary of generic causes of track quality deterioration

Recommendations for treatment – Wheelock Embankment

Following this preliminary interpretation, recommendations were given for future FWD testing at locations throughout the route, where poor stiffness and/or critical velocity was deemed to be a cause for concern. One such site was chosen, which comprised 1mile 60chains where the track was generally on embankment ranging in height from 1m to 15m. Line speed was limited to 60mph, but it was hoped to increase this to 100mph. To the south, the embankment is underlain by Boulder Clay, but moving further north the underlying geology becomes variable and consists of Glacial Sand and Gravel, Alluvium, Boulder Clay and Fluvio Glacial Deposits.

The site was broken down into 12no. sections of similar track quality, trackbed condition (from the ABS), FWD stiffness and Critical Velocity values. Figure 5 shows a schematic of the track (at the top) for Sections 3-6, including features observed during the Walkover Survey. The FWD stiffness profile (Sleeper Support Stiffness) is shown (at the bottom) with a dotted line representing the minimum value for a new track of 60kN/mm/sleeper end.

Below this is a plot of the estimated Critical Velocity and 70% of that value averaged over 80m. Individual 70% values are not plotted because this could give the occasional isolated low value that would not, in itself, result in 'Dynamic Interaction'. It is considered that a single low value does not necessarily limit the line speed to that value, as the FWD measures only the velocity of wave propagation locally to the test position, i.e. it is representative of a length of track of less than 2m. The dotted line shown on this profile represents a line speed of 125mph.

Results were presented on Total Route Evaluation Plots, each representing half a mile, and Treatment Codes were developed to summarise the maintenance and renewals options for each section.

On WCML Route 7, the results from the Preliminary Phase of Total Route Evaluation were used to identify sections where there were concerns regarding inadequate stiffness, or there was the potential for dynamic interaction that would limit any proposed increase in line speeds.

FWD testing was used to measure Sleeper Support Stiffness and estimate the Critical Velocity. These results were used in conjunction with all the other information obtained during Total Route Evaluation to develop specific recommendations to improve stiffness and/or Critical Velocity; whilst also ensuring that a trackbed would be constructed that offered the potential for low maintenance over a complex site.

References

Armitage, RJ, Brough, MJ, Sharpe, P, *Total Route Evaluation on ECML Freight Primary Route 3*, Railway Engineering 2003, 6th International Conference and Exhibition, London, 30April – 1 May 2003

BS1377: Part 2, *British Standard Methods of Test of Soils for Civil Engineering Purposes* , Classification Tests, 1990,

Collop, AC, Grainger, PS, Sharpe, P, *Predicting the effect of Trackbed Stiffness on Track Quality*, Railway Engineering 2001, 4th International Conference and Exhibition, London, 30April – 1 May 2001

Hunt, GA, *Analysis of Requirements for Railway Construction on Soft Ground*, LR TM 031, British Rail Research, Derby, 1993

Hunt, GA, *Eurobalt II Final Report to Railtrack*: Optimisation of Ballasted Track, RR-TCE-2000-311, AEA Technology Rail, September 2000

RT/CE/S/006, *Track Ballast and Stoneblower Aggregate* (Issue 3, August 2000)

RT/LS/S/011, *Prevention of Damage to and Danger from Surface & Buried Services* (Issue 3, February 2002)

RT/CE/C/039, *Code of Practice for Formation Treatments* (Draft 1d, January 2003)

RT/CE/S/102, *Track Construction Standards* (Issue 5, February 2002)

RT/CE/S/104, *Track Maintenance Requirements* (Issue 5, June 2000)

Sharpe, P, *Formation Assessment for Maintenance and Renewals Planning – Total Route Evaluation*, Railway Engineering 1999, 2nd International Conference and Exhibition, London, 25-26 May 1999

Sharpe, P, *Trackbed Investigation, Journal and Report of Proceedings*, The Permanent Way Institution, Vol. 118, Part 3, pp238-255, 2000

Sharpe, P, and Grainger, PS, *The Preliminary Investigation Phase in Total Route Evaluation*, Railway Engineering 2000, 3rd International Conference and Exhibition, London, 5-6 July 2000

Detecting voids beneath concrete pavements using surface deflection

D. Cudworth
Babtie Pavement Management and Engineering

Introduction

Jointed concrete pavements have been used in the UK for both highway pavements and industrial ground floor slabs for over eighty years. This paper discusses the principal long term maintenance concern with this type of pavement construction, namely the joints. It concentrates on the problem of under-slab voiding at joint locations and outlines a methodology that can be used to detect voids using deflection testing.

The paper draws largely on the experiences of the author, who since 1994 has worked with BALVAC to develop a testing methodology for the identification of under-slab voids to target repairs using the vacuum grouting technique.

The history of jointed concrete pavements in the UK

The rise and fall of jointed concrete road pavements

An access to a goods yard at Inverness railway station constructed in 1865 was probably the first recorded use of concrete as a running surface in the UK (Maxwell, 1992). The pavement, which was 40m in length and 175mm thick, was the forerunner to a more heavily trafficked trial site in Edinburgh which was constructed in the same year. The trials were conducted under the guidance of Joseph Mitchell and initially provided several years of good service until failure occurred during a severe winter. Although the pavements almost certainly failed as a result of frost damage it was concluded that concrete became too brittle with age to be used as a running surface.

Whilst the first successful concrete pavements were probably constructed in the USA in the 1890's (Huang, 1993), concrete was not generally used for road construction in the UK until the 1920's. By this time there was a better understanding of concrete design and quality control and it was no longer considered to be too brittle to be used. Volume six of the Roadmakers Library

Transportation geotechnics. Thomas Telford, London, 2003

(Collins and Hart 1936), states "A most interesting feature of modern road development is undoubtedly the rapid increase in the use of concrete. In 1926, the total mileage of concrete roads in the British Isles was 182, and when this figure is compared with the total of some 3,000 miles in 1935, it is probable that the rapidity of increase is due to the inherent advantages which this mixture has over other types". In 1955 the first formal design guide for jointed highway pavements was published, (Road Note 19, 1955). This new document also incorporated some features developed during the construction of the German Autobahn's during the 1930s. The principal problem with jointed concrete pavements is the joints between the slabs, these discontinuities in the pavement structure are prone to deterioration and are often difficult to maintain.

Despite the early optimism for the use of concrete, by the end of the century jointed concrete highway pavements had fallen from favour and for large road construction projects they have been largely replaced with joint less continuously reinforced concrete slabs overlaid with bituminous surfacing. In an attempt to improve the ride and acoustic properties of existing jointed concrete pavements the Highways Agency (HA) has instigated a policy to overlay all their concrete roads. Consequently, the structure of many concrete pavements are currently being transformed using a crack and seating technique and/or being buried under bituminous overlays.

New jointed concrete construction in the UK

Despite jointed concrete construction being virtually abandoned by highway authorities, this has had very little influence on the area of jointed concrete pavements constructed in the UK. Annually the flooring industry constructs approximately 6000000 m^2 (1000000 m^3) of new pavements, which is approximately an equivalent area to 300 km of a dual, three lane highway.

For industrial ground floor slabs in particular, joints are still a significant problem and in recent years the flooring industry has experimented with large bay construction to minimise their use. The dilemma is that to achieve the flatness criteria demanded by operators, (who want to use narrow aisles and high racking), a robust concrete surface is required and where this contains joints the required flatness may not be achieved due to curling and/or movement (deflection) across joints related to foundation support.

Why is under-slab voiding a problem

The purpose of a pavement is to transmit the wheel loads imposed on the structure into the underlying subgrade (via the foundation consisting of sub-base and/or capping). Voids beneath the base of the slab and its foundation can seriously reduce the performance of the pavement, which can ultimately result in premature failure. The way in which failure occurs and its influence on the pavement user largely depends upon the type of pavement, these can generally be divided into two main groups, internal (floor slabs) and external (roads and paving).

Highways and other external pavements

There are a number of reasons why joints can cause problems. In a highway pavement, failure of the joint sealant is often followed by rainwater ingress into the pavement structure. Under repeated loading the joints will deteriorate and ultimately this weakening can result in voids developing between the base of the slab and the underlying foundation.

Since the fundamental requirement of a slab is to spread the imposed wheel loads evenly throughout the pavement foundation, any voiding however small, can have a serious influence on performance. When a void develops beneath a joint in a highway pavement the slabs are more likely to crack due to fatigue and/or settlement. This in turn will allow further rainwater ingress into the foundation, increasing the deterioration and ultimately leads to total pavement failure. Voids also increase the vertical deflection at the edges of the slab, which promotes a further reduction in the performance of the joint. Rainwater ingress together with large vertical movement can cause pumping in the foundation and in the most extreme cases can result in the shearing of the joint dowel bars. When the physical connection between the slabs is broken differential settlement is more likely to occur across the joint which subsequently produces further faulting.

Industrial ground floor slabs

Rainwater ingress is not a concern for industrial ground floor slabs, however, voiding can be a significant issue due to the character of the Mechanical Handling Equipment (MHE) using the pavement. Whereas highways and other external pavements are generally trafficked by vehicles with pneumatic tyres and some form of suspension, most warehouse MHE, (such as that shown in Photograph 1), have hard polyurethane tyres and no suspension.

Similar to highway pavements, voids result in increased vertical deflection at joints. However, due to pneumatic tyres and suspension motorists are usually unaware of poor joints until the road is closed for maintenance.

In contrast, the MHE operator, driving a vehicle with no suspension and hard wheels, notices any small variation in the surface of a slab. Joints with voiding are easily identifiable by the operative through a process which was graphically described by Pearson (1999) as "joint bounce" where the vehicle bounces and rocks over the joint.

In the least extreme cases a MHE truck 'bouncing' over a voided joint is uncomfortable, especially where a "reach" truck operator is working at a height of 12m (and hence the movements are magnified). In the most extreme cases 'joint bounce' has resulted in serious Heath and Safety concerns and it is not unusual for a warehouse risk assessment to identity immediate remedial treatment.

Photograph 1, Typical warehouse mechanical handling plant

Why do voids develop beneath joints?

There are several causes of under-slab voiding but typically they are caused by poor construction and/or deterioration following repeated traffic loading.

Slab curling and poor construction

Inadequate compaction or settlement of the foundation has been known to produce voiding even before the pavement is subjected to traffic loading. This problem is more typical for floor slabs than for highway pavements since during road construction more attention is given to the quality of capping and sub-base layers. The author is aware of at least one site where the contractor maintained that a floor slab was built on a good quality sub-base whilst there was evidence of deformation of the sub-base layer by delivery trucks during construction.

In addition to poor foundation layers, voids can develop between the base of the slab and the underlying foundation by joint curling. If as the slab cures the surface shrinks more than the base, the edges of the slab can curl upwards causing a void, this is relatively common in floor construction but is probably less common in thicker highway slabs.

Traffic induced voiding (asymmetric loading)

Voids can develop under concrete pavements, which may well have been adequately constructed, by a process known as asymmetric loading condition. As a wheel moves forward towards a discontinuity such as a joint (or even a crack) the pavement will gradually deflect. With reference to the direction of the moving wheel, the slab upstream of the joint is referred to as the approach slab (or approach side of joint) and the downstream slab the leave. This is shown schematically in Figure 1.

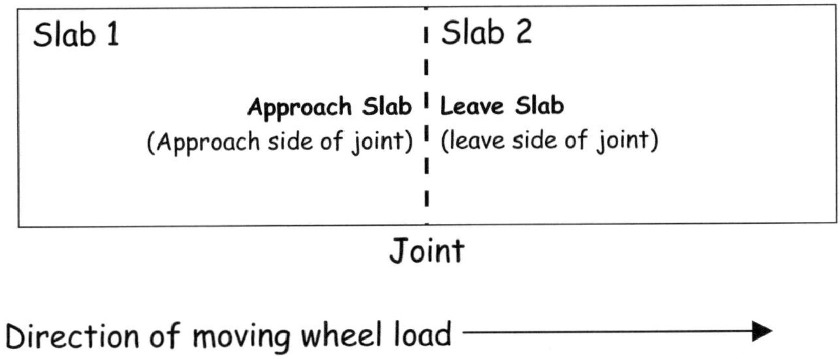

Figure 1, Slab approach and leave definition

As the moving wheel approaches a joint there will be a gradual loading of the pavement at the slab approach, as shown schematically in Figure 2.

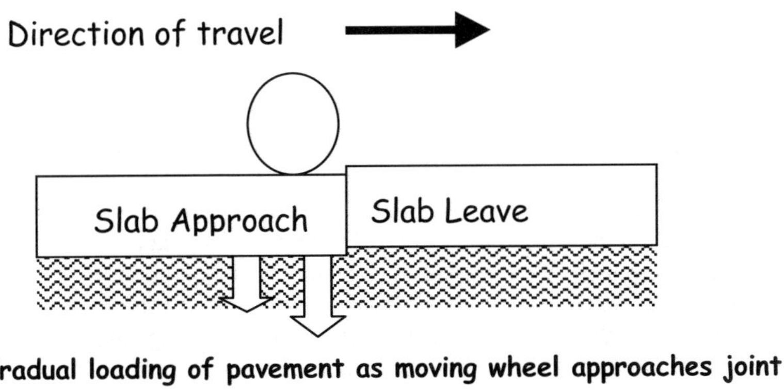

Figure 2, Approach loading condition

At a point immediately before the wheel arrives at the joint, there will be a step between the loaded approach slab and the adjacent unloaded leave slab. As the moving wheel impacts against the step there will be a sudden loading at the leave side.

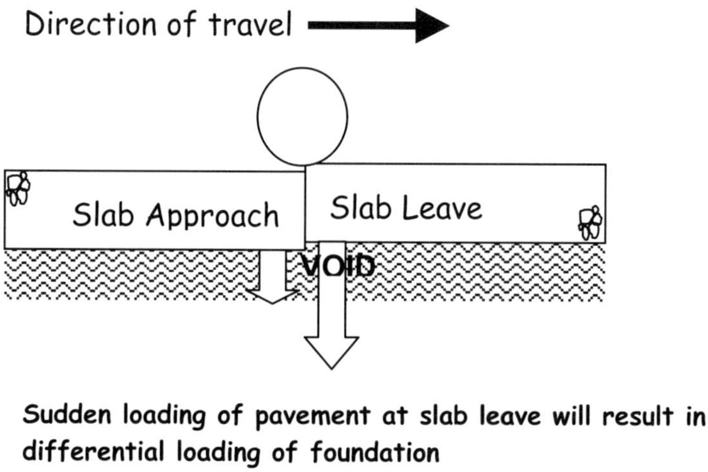

Figure 3, Leave loading condition

Where traffic flows in one direction only, voids caused by asymmetric loading develop on the leave side of the joint and propagate against the flow of traffic towards the approach side.

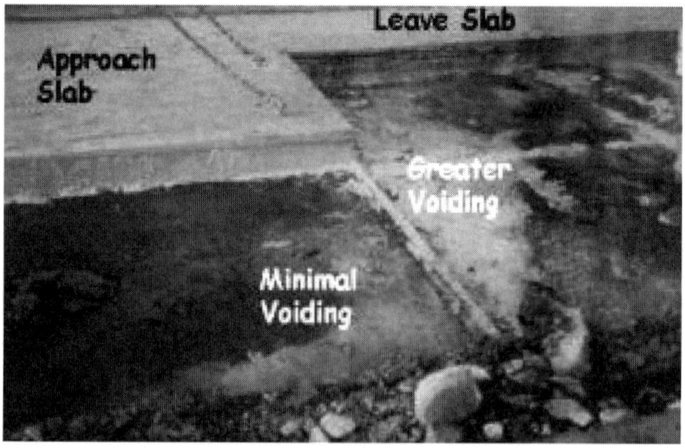

Photograph 2, Slab lifting to show resin injection under voided joints

This hypothesis was proven by BALVAC in 1994 when a joint was injected with coloured resin and when the slabs were lifted more resin was observed under the leave slab (Photograph 2).

Pavement testing equipment

To allow an assessment of the formation of voiding beneath slabs pavement deflection testing can be used. During the 1960's the Transport and Road Research Laboratory (TRRL) undertook a programme of trials to monitor the structural condition of flexible pavements using Benkelman Beams. The Benkelman Beam, was developed at the WASHO road test, and is a simple and inexpensive apparatus for measuring pavement deflection between the dual tyres of a 63.5kN axle.

Based on their research the TRRL were able to link "standard pavement deflections" to pavement structural condition for flexible pavements. The UK specification for Benkelman Beam testing was first published in 1978 (Kennedy et al, 1978). The analysis procedure is entirely empirical and therefore only reliable for evaluation of pavements similar to those for which the empirical relationships were developed. Consequently, in the UK, the Blenkelman Beam was not widely used for the evaluation of concrete pavements and is unable to be used for void detection.

During the 1970's an automated Benkelman Beam, the Lacroix Deflectograph, was developed in France. In the late 1970's the TRRL and a road testing equipment manufacturer WDM of Bristol, developed the Deflectograph for use in the UK. Apart from an improved survey speed this automated beam also measured deflections in both wheel paths.

During a Deflectograph survey pavement deflections are measured as the rear wheel of the Deflectograph approaches the tip of a lever arm, which is attached to a measuring head. Once the deflection measurement is recorded a winch is activated which drags the measuring frame forward at twice the speed of the moving vehicle to the next measurement point. This sequence is repeated as the vehicle travels along the road at a speed of approximately 2km/h. Unlike the Benkelman Beam Survey the Deflectograph testing process is entirely automated.

During the 1980's WDM modified a Deflectograph to enable deflection measurements to be taken in the nearside wheel path on either side of a joint between two slabs. To be able to make deflection measurements at joints, the following modifications were made to the standard Deflectograph (Photograph 3):

- The addition of a second measuring head to the nearside of the reference frame to allow deflections to be measured on both sides of a joint.

- The extension of the reference frame to move the deflection beam outside the line of the rear wheels.

- The introduction of a procedure (frame lock) to interrupt the normal continuous nature of the measuring system and hold the measuring frame in its forward position until the specific joint location is reached.

Photograph 3, Modified Deflectograph for joint testing

The normal sequence of recording deflections at fixed intervals is interrupted using the frame lock device. During testing the frame is positioned with measuring beams either side of the joint by an operator who walks alongside the equipment. When the measuring heads are located at the correct position, the frame-lock is released (by the operator) and the Deflectograph follows its normal sequence. At the end of the sequence the frame is pulled forward to its starting position where the frame is locked , ready for the next joint.

At each joint location, as the rolling wheel moves towards it the deflections at the tips of the measuring beams on the surface of the concrete slab are measured every 10mm, subsequently two deflection bowls are measured. At each joint 135 individual measurements are recorded. The shapes of each pair of deflection bowls (either side of the joint) are used to provide an indication of the degree of load transfer at the joint. Deflection bowls with similar shapes indicate good load transfer, whereas poor load transfer results in bowls of dissimilar shape. The Deflectograph also measures the peak deflections either side of the joint.

The principal disadvantages with these measurements are that the feet of the reference frame lie within the zone of influence of the loaded wheels. Consequently the weight of the equipment has a significant influence on the measured deflection.

Whilst the Modified Deflectograph provides an effective method of assessing the overall condition of joints, the measured deflections are not sufficiently accurate to enable the detailed analysis needed to predict small under-slab voids.

Development of a deflection testing methodology to detect voids

During the late 1970's BALVAC perfected a vacuum grouting technique for the injection of under-slab voids beneath jointed concrete pavements. In order to assess the need for remedial treatment and to demonstrate the effectiveness of the process it was realised pavement testing would be required. BALVAC utilised deflection testing and following a period of development in 1980s Department Advice Note HA/6/80 *Vacuum Grouting of Concrete Road Slabs* was published by The Department of Transport. (HA 6/80 1980).

The testing methodology outlined in HA 6/80 is based on measurement of an absolute joint deflection under a 6350kg static load. Deflections are measured using long travel dial gauges positioned 100mm from the edge of a joint. If absolute deflection of the slab exceeded 0.4mm or the relative deflection (deflection difference) between measurements taken on either side of a joint exceed 0.2mm then remedial treatment should be considered.

Unlike the modified Deflectograph the testing methodology was developed specifically for the detection of under-slab voiding. (Photograph 4).

Photograph 4, HA 6/80 dial gauge testing

Limitations of modified Deflectograph and HA 6/80 testing

The modified Deflectograph was automated and produced relatively accurate deflection measurements. However, the deflection bowls obtained from the survey were not considered to be sufficiently high resolution to detect under-slab voids. In contrast, the methodology developed in HA 6/80 was specifically produced to identify joints with voiding problems, however, the survey was slow and accuracy of the data was questionable.

In order to develop a reliable method for predicting under-slab voids it was realised that the pavement testing equipment required the following specifications:

- Ideally it must produce a load pulse with the magnitude, shape and duration similar to a moving wheel load and needs to be measured accurately,
- It must be automated and measure deflection measurements to an accuracy of around 0.001mm (1 micron),
- In order to measure true absolute slab deflections the self weight of the equipment should have little or no influence on the deflection measurements.

Accurate deflection and load measurements, which could be used to reliably detect voiding only became available following the introduction of the Falling Weight Deflectometer into the UK in the late 1980's.

The Falling Weight Deflectometer (FWD)

The most versatile and accurate testing equipment currently available for measuring pavement surface deflection is the Falling Weight Deflectometer (FWD). Since the late 1980's FWD pavement testing has become more widely available in the UK. The FWD (Photograph 5) closely simulates realistic heavy wheel loads necessary for testing the structural condition of a pavement. The force pulse is obtained by dropping a weight on a specially designed spring system. The equipment is trailer mounted and is capable of testing in the towing position by automatically lowering and raising the loading plate and seismic detector bar to and from the pavement surface whilst the trailer is stationary. Automated controls within the towing vehicle allow any selected sequence of up to 64 drops of the falling weight from any combination of 4 drop heights.

With the loading plate on the pavement surface the full energy of a single falling mass with a peak load magnitude of 30kN to 240kN is transferred through rubber buffers to the loading plate. The configuration of the buffers is such that for different falling masses from any of the four drop heights a loading time of between 25 and 30 msec is achieved with a rise time (maximum load), of between 10 and 15 msec. The loading sub-assembly (buffers and loading plate) and the falling weight is operated by hydraulic cylinders which are pressurised by a hydraulic pump driven by an electric motor.

Photograph 5, Falling Weight Deflectometer during joint testing

The equipment is provided with two different sizes of interchangeable loading plate (300mm and 450mm diameter). A load cell, which is situated directly above plate, measures the load, which is applied perpendicular to the loading plate, to an accuracy of at least 1kPa. Pavement surface deflections are measured in microns (m x 10^{-6}) by seismic velocity transducers, (geophones). A single geophone is situated in the centre of the loading plate directly under the load. The remaining geophones are positioned in holders along the raise/lower bar at radial distances of up to 2.45m from the centre of the load. The geophone holders are spring loaded, which ensures a good contact between the geophone and the pavement surface. Additionally tests can also be undertaken using a rear geophone extension bar, which allows geophones to be positioned up to 0.45m rearward from the centre of the load (Photograph 6)

Photograph 6, FWD geophones positioned across a joint behind the load plate

The significance of the FWD testing equipment on void detection

The importance of the FWD in the advancement of pavement testing in the UK cannot be underestimated. For the first time it allowed pavement engineers to easily measure very accurate load and deflection measurements. In addition to its accuracy, the ability to test at different load levels provides the necessary data for accurate void detection. The FWD has replaced all other methods of deflection testing for the detailed evaluation of pavement structures.

M25 vacuum grouting project junction 5 to county boundary

In 1995 the largest ever UK grouting project to eliminate voids beneath a jointed concrete pavement was proposed. During the project over 44000m^2 of pavement were to be treated in a limited working time that required six injection teams to work concurrently. The project required new levels of productivity to be achieved for the technique and since joint testing was required both before and after injection it was decided that HA 6/80 testing would not be feasible. Consequently, due to its accuracy and speed FWD testing was proposed.

In deciding on the testing methodology it was assumed that the FWD would be used to replicate HA 6/80. Consequently, the first step in developing a FWD testing methodology was to establish the FWD load which would be approximately equivalent to the static loading used during HA 6/80 surveys. Based on an analytical model of the pavement it was established that a 50 kN FWD load and 300mm diameter loading plate would be similar to the 6350kg static axle load used in HA 6/80. The proposed test criteria was also similar to HA 6/80, however, in addition to absolute deflection and relative deflection measurements, joint load transfer efficiency (ratio of deflection across a joint) and void intercept values (defined below) were also calculated.

Absolute Deflection

Whilst the absolute deflection from the FWD survey was broadly similar to HA 6/80 it was not possible to exactly replicate the survey. In HA 6/80 the dial gauges are positioned 100mm either side of the joint, this is not possible during a FWD survey since the radius of the smaller loading plate is 150mm. The FWD absolute deflection is therefore calculated 250mm from the joint using the geophone positioned at the centre of the loading plate. Deflections on either side of the joint are measured using geophones positioned 200 and 300 mm from the centre of the load. The testing configuration used for the M25 survey, which was also adopted for subsequent projects is shown in Figure 4.

In order to determine any difference between slab approach and leave conditions two tests were undertaken at each joint location. During the first test the FWD loading plate was positioned on the approach side of the joint. For the second test the testing equipment was repositioned on the leave side and the joint deflection measured using geophones positioned 200 and 300mm behind the loading plate.

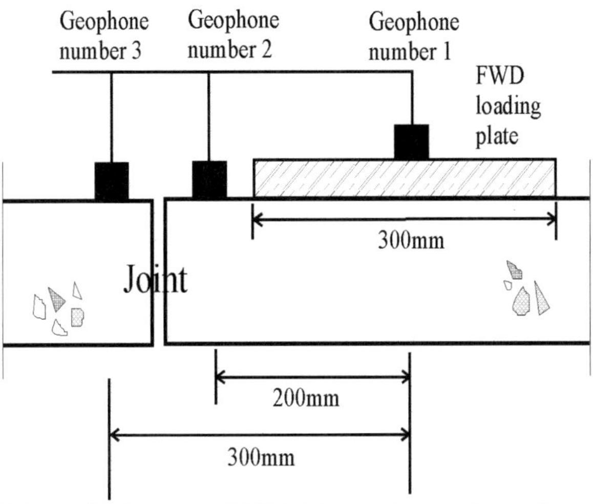

Figure 4, Schematic diagram of FWD joint testing configuration

During HA 6/80 testing a joint with greater than 400 microns deflection was assumed to require treatment. An analysis of the FWD deflections indicated that an intervention level of 240 microns should be used for FWD analysis. Consequently, any joint with a deflection greater than 240 microns was identified for grouting and following injection it was retested to confirm that the post grout deflection was below the treatment intervention level.

Relative deflection

HA 6/80 presented guidance for assessing relative deflection, which suggested that values exceeding 200 microns should be considered for remedial treatment. Due to the increased accuracy of the FWD tests it was decided that this criteria could be reduced to 100 microns,

Joint load transfer efficiency

For load transfer efficiency (LT) calculations the following equation, from the Design Manual for Roads and Bridges (HD29/94), was used:

$$LT\,\% = \frac{\text{Deflection of unloaded slab}}{\text{Deflection of loaded slab}} \times 100 \qquad (1)$$

Since the vacuum grouting process does not improve the physical bond between slabs load transfer was not initially considered to be a significant parameter for assessment of treatment.

Void intercept analysis

Although an analysis procedure known as void intercept analysis was becoming increasingly popular, especially in the USA, it is not included in HD29/94. Detailed guidance was however published by the American Association of State Highway and Transportation Officials (AASHTO, 1993) and was adopted during the M25 project.

To determine void intercept values at least three FWD tests were carried out at different load levels at each joint test location. Voids underneath the slab are closed as the pavement is loaded. Increasing the applied load further increases the closure of the voids resulting in a deflection that will not improve proportional with load level. By extrapolating from the three tests the deflection corresponding to zero loading is calculated by liner regression, as shown in Figure 5.

Theoretically for zero load the corresponding pavement deflection should be approximately zero. However, for a voided foundation, where deflections may not increase proportionally with load, a linear regression analysis may not intercept the origin of the load versus deflection plot. The point at which the linear regression line intercepts the y-axis is known as the void intercept.

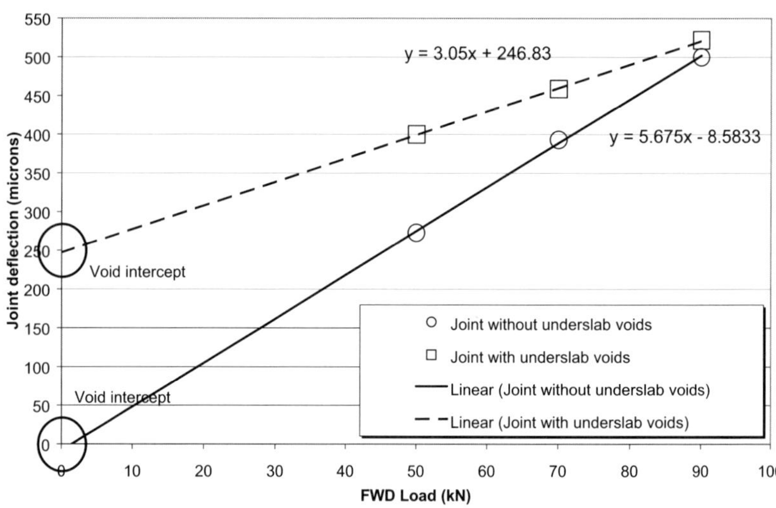

Figure 5, Void intercept analysis

In the AASHTO guidance void intercept values greater than 50 microns are suggested as being indicative of under-slab voiding. This criteria was adopted for the M25 project.

M25 summary of FWD testing results

A statistical comparison of slab leave and approach joint performance was undertaken using over ten thousand FWD measurements (5400 joints) from the pavement evaluation testing for the project. The results of the Pavement testing (normalised to a 50 kN load) are presented in Tables 1 to 3.

The results of the statistical comparison of leave and approach tests confirmed that asymmetric loading at a joint location was more likely to cause deterioration of the foundation beneath the leave end of the slab. Based on 85th percentile values, before grouting, the leave deflection was found to be 28% higher than the approach. Following grouting, when the voiding had been eliminated, deflections on either side of the joint were found to be identical. Moreover, the 85th percentile values, which are usually used for pavement evaluation purposes, were below the 240 micron intervention criteria. Based on 50th percentile values, vacuum grouting resulted in load transfer efficiency increasing from 41 to 75%, this was unexpected and was due to the filling of the voids.

The void intercept values prior to grouting confirmed that voiding was significantly more likely beneath the leave side of the slab than the approach, again consistent with under-slab voiding caused by asymmetric loading.

The absolute deflection and void intercept results obtained from the M25 survey were significant in that for the first time deflection testing had been used to reliably detect voids and prove the hypothesis of asymmetric loading.

Table 1, Slab leave and approach absolute deflections

Percentile	FWD results before grouting		FWD results after grouting	
	Approach (microns)	Leave (microns)	Approach (microns)	Leave (microns)
10th	110	110	94	94
20th	126	126	110	110
30th	142	142	126	110
40th	157	173	126	126
50th	189	220	142	142
60th	220	268	157	157
70th	236	299	189	173
85th	**283**	**362**	**236**	**236**

Table 2, Slab leave and approach joint load transfers

Percentile	FWD results before grouting		FWD results after grouting	
	Approach (microns)	Leave (microns)	Approach (microns)	Leave (microns)
10th	19	13	25	25
20th	21	16	33	36
30th	24	20	33	36
40th	29	27	62	65
50th	**41**	**39**	**75**	**77**
60th	69	64	83	83
70th	84	80	87	87
85th	90	98	93	92

Table 3, Slab leave and approach void intercept values

Percentile	FWD results before grouting		FWD results after grouting	
	Approach (microns)	Leave (microns)	Approach (microns)	Leave (microns)
30th	10	10	<0	<0
40th	10	10	10	<0
50th	10	20	10	10
60th	10	20	10	10
70th	10	30	10	10
80th	20	40	20	20
85th	**20**	**50**	**20**	**20**
90th	20	70	20	30

The significance of the M25 project

The M25 project was a significant for the following reasons:

- It demonstrated that FWD testing was an accurate and efficient method of detecting under-slab voids (up to 30 joints could be tested in 1 hour).

- The results could be used to prove asymmetric loading was a cause of under slabs voiding.

- The significance of foundation support on load transfer efficiency could be measured.

- It demonstrated that void intercept analysis could be used to detect voids. Furthermore, the results also indicated that an intervention criteria of 25 microns could be more appropriate than the 50 micron criteria proposed by AASHTO.

Testing and analysis developments following the M25

Following the M25 project the FWD has continued to be used successfully for pre and post vacuum grout treatment testing. These experiences have resulted in improvements in testing and analysis methodologies.

A better understanding of absolute deflection

For the M25 project it was assumed that a single absolute deflection criteria could be applied to the entire site. Furthermore, HA 6/80 proposed single criteria for all concrete joint testing. In reality the absolute deflection of the loaded slab will be a function of the concrete stiffness, foundation support, slab thickness, wheel load and joint load transfer efficiency. The critical parameters which influence the deflection at a joint or crack produced by dynamic loading are shown in Figure 6.

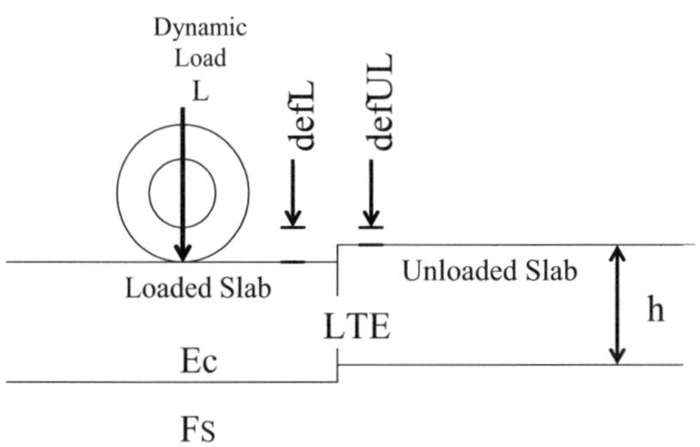

Where:

h = slab thickness (mm)
E_c = concrete stiffness (MPa)
F_s = foundation support (MPa or k-value)
L = wheel load (kN)
LTE = load transfer efficiency between slabs (%)
defUL = deflection of unloaded slab (microns)
defL = deflection of loaded slab (microns)

Figure 6, Critical parameters that influence absolute deflection

Using analytical pavement models incorporating the parameters shown in Figure 6 and previous FWD test data where available, it has been possible to predict the influence of load transfer efficiency on absolute deflection. Based on this approach absolute deflection criteria have been developed for individual sites and for the load transfer efficiency of the joint under consideration. This has resulted in a more discriminative analysis based on site and even joint specific criteria.

Void intercept criteria

Experience following the initial M25 survey has generally supported the finding that void intercept analysis provides a reliable method of predicting voids. Whilst larger void intercepts generally indicate bigger voids it has not been possible to produce a correlation between intercept values and the volume of resin injected into the joint during treatment. This can be explained when considering two voids of identical volume but different shapes. Each joint will have the same resin absorption but different deflections due to the different void shapes.

Relative deflection

Due to the accuracy of the FWD it is possible to consider deflections that are considerably smaller than those measured during HA 6/80 testing. Consequently, experience has demonstrated that the 200 micron HA 6/80 relative deflection criteria was inappropriate, based on FWD testing a value of 50 microns is typically assumed. This is especially relevant if the pavement is to be overlaid with bituminous material, i.e. where reflective cracking is undesirable. Furthermore, this value, which is based on experience gained in the UK is consistent with guidance presented by the American Asphalt Institute, (1983).

Industrial ground floor slabs

In recent years the methodology void detection developed for highway pavements has been successfully adapted for the evaluation of industrial ground floor slabs. Falling Weight Deflectometer testing is especially appropriate for testing warehouse pavements since the equipment is able to measure the small deflections that can have a very significant influence on the performance of the floor.

During an assessment of a large warehouse facility it was possible to correlate the visual deterioration of the joint with the step between slabs measured by the FWD. This survey demonstrated that even small increases in joint deflection can cause significant increases in joint damage, especially where joints are trafficked by MHE with hard polyurethane wheels. Industrial ground floor slab surveys have also demonstrated voiding caused by slab curling and poor foundation support.

Photograph 7, FWD testing of an industrial ground floor slab

Future Developments

The experience gained from nearly ten years of FWD surveys in conjunction with vacuum grouting has been extremely successful and has resulted in a reliable method for identifying voids. However, there are a number of issues which require further investigation, namely, the effects of loading time and pavement temperature. The author has undertaken some tentative analysis of these factors, the results of which are presented below.

The effect of loading time on slab deflections

The FWD loading configuration is such that in its normal test set-up the load pulse is approximately equivalent to a 30 kph dynamic load (0.03 seconds). Different types of FWD or different types of deflection testing equipment may have significantly different loading times. For example, the modified Deflectograph loads the pavement at around 2.5 kph whereas the HA/6/80 load is static. Although the stiffness of concrete is not affected by loading time (as a bituminous material would be), it has been suggested that a slab will react differently to changes in loading time.

To investigate the effect of loading time on the response of a slab the FWD was positioned with the centre of the loading plate 250 mm from a joint. A number of tests were undertaken to assess whether voiding was present beneath the joint. Once it had been established that voids were not present a series of tests were undertaken during which the rubber buffers onto which the weights fall were successively removed. Decreasing the number of buffers from ten (standard set-up) to four increases the loading time of the FWD from 0.03 seconds to 0.05 seconds (equivalent to a dynamic loading range in terms of speed of 25 to 30 kph). The effect of the different loading times on the maximum deflection under a 40 kN load (300mm diameter loading plate) is shown in Table 4.

Table 4, The effect of loading time on slab deflection

Loading time (Seconds)	Approximate equivalent speed of load (km/hour)	Deflection under FWD load plate (40 kN load) (microns)
0.030	30	167
0.035	28	169
0.040	25	171
0.045	22	174
0.050	20	176

Based on the results in Table 4 the following relationship was determined between the deflection under the FWD plate (40 kN load) and the Loading time of the pavement (L_T).

$$\textit{FWD deflection (40 kN load)} = 420.91\, L_T + 154.61 \qquad (2)$$

Where: L_T is in seconds

Equation 2 was only determined based on deflection measurements for FWD loads of 40 kN and pavement loading times of between 0.03 and 0.05 seconds. Assuming it can be used for longer loading times the equation suggests that slow moving wheel loads may produce significantly higher maximum deflections than fast moving wheel loads. Although equation 2 is based on a very limited amount of data from only one test site it does indicate that the maximum deflection of a concrete slab could in part be dependent upon loading time (vehicle speed). Subsequently, for two identical magnitudes of load, two identical slabs may react differently if the load pulses are not the same. Unfortunately, without further testing it is not possible to assess the significance of loading time on slab reaction, and in reality small differences due to loading time may be insignificant when the dynamic effects of heavy vehicles "bouncing" over joints are considered.

Seasonal and temperature variations in joint efficiency

Research has shown that the efficiency of dowel bars is significantly effected by pavement temperature and temperature gradients throughout the slab (Geer 1990 and Larsen 1990). To investigate the effects of pavement and air temperatures on joint performance efficiency a trial section of pavement consisting of five joints was monitored at different temperatures using a FWD. During construction of the pavement thermocouples were inserted at the surface and base of one slab to enable measurements of temperature gradients to be undertaken. The slabs were constructed by hand using C40 air entrained concrete with two layers of mesh reinforcement. All five joints were expansion joints constructed using 25mm diameter steel dowel bars. The sub-base consisted of a Type 1 granular material, it was noted that during construction there was no separation membrane between the base of the slab and its foundation. Traffic flow along the trial section was restricted to a few cars each day and no commercial vehicles. The site was constructed on a gradient and there was no surface water drainage. Testing of the pavement was undertaken on four occasions one autumn.

The joint comparison (Table 5) confirms the other research referred to above and indicates that the joint load transfer efficiency is significantly increased at high temperatures. During the testing undertaken on 21st November it was noted that ice had formed in the joint groove (night-time temperature had been well below 0°C) and this appears to have a similar "joint locking" effect to high temperatures.

Table 5, Comparison of joint load transfer efficiency at different temperatures

Date of FWD testing	Temperature (degrees C)			Mean joint load transfer (%)	Mean absolute deflection (microns)
	Surface of slab	Bottom of slab	Air		
18/08	33	23	30	83	286
23/08	18	16	15	41	487
12/10	9	8	9	27	363
21/11	0	0	0	74	262

Conclusions and future work

It is recognised internationally that the detailed investigations made possible by FWD technology produce versatile and reliable solutions to analysing all pavement structures. The detailed analysis methodologies, which are only made possible due to the accuracy and repeatability of the testing equipment, has facilitated a thorough and meticulous approach to forensic investigation of pavements.

For jointed concrete pavements the current FWD methods outlined in this paper have clearly led to greater reliability in quantifying joint performance. The significance of being able to accurately measure pavement deflection is that overall influence of under-slab voiding on the structure of the pavement can be quantified. Consequently, it has also been possible to prove the hypothesis of asymmetric loading being a contributory factor in void development and propagation.

As with all investigation processes great care is needed in the interpretation of the results. Since no two concrete slabs are alike it has been demonstrated that it may be necessary to develop site specific criteria for void identification.

Further research is clearly required to confirm the influence of pavement temperature and loading time on the performance of joints. It would be prudent to include in these surveys other areas of investigation such as ground penetrating radar to provide correlation with the other techniques. The methodology developed could potentially be further developed to enable assessment of void size and distribution.

References

AASHTO. *Design of pavement structures*. Published by the American Association of State Highway and Transportation Officials (AASHTO), 1993.

Asphalt Institute *Asphalt overlays for highway and street rehabilitation. 1983*

Collins H.J. and Hart C.A. *The principals of road engineering. volume 6 the roadmakers library,* Arnold, London 1936.

Department of Transport. *Vacuum grouting of concrete road slabs,* Department of Transport Departmental Advice Note HA 6/80, 1980.

HD29/94 Department of Transport. *Structural assessment methods*. HD29 Design Manual for Roads and Bridges, DMRB 7.2.3. 1994.

Geer W.C. *Seasonal variation in joint efficiency of dowelled concrete pavements*. Proceedings from the 3rd International Conference, Bearing Capacity of Roads and Airfields, Norway, 1990.

Huang. *Pavement analysis and design*. Prentice – Hall, New Jersey, 1993.

Kennedy C.K. Fevre P. and Clarke C.S. *Pavement deflections: equipment for measurement in the U.K.* TRRL Report LR834, 1978.

Larsen T. *Test procedure for rigid pavements using falling weight deflectometer*. Proceedings from the 3rd International Conference, Bearing Capacity of Roads and Airfields, Norway, 1990.

Maxwell G. *Ways of the world.* Rutgers University Press, Australia 1992.

Pearson D. *Proceedings of the 3rd European symposium on the performance of bituminous and hydraulic materials in pavement, 1999.*

Road Research Laboratory. *Road note 19, a guide to the structural design of pavement for roads. Concrete roads, design and construction*. Her Majesty's Stationary Office, London 1955.

Private finance initiative infrastructure projects – implications on geotechnical design

S. Jarvis,
Arup
P. Gilbert,
Atkins

The views expressed in this paper are those of the Authors and do not necessarily represent the views of their respective organisations.

Introduction

Today, most highway projects are procured on a 'Design and Build' (D&B) basis, occasionally within a 'Private Finance Initiative' (PFI) framework. Designers now have a wealth of experience in this area, however a cautious approach is required because of risks inherent within the contractual framework. Some Contractors have experienced financial difficulties because of major differences between final outturn costs and the costs expressed at tender. Designers therefore need to be extremely careful in their approach to working for a Contractor Client.

This paper describes some of the issues that the Authors have experienced on a number of significant D&B projects with particular emphasis on the geotechnical design. Each stage of the project is described and the lessons learned and observations made from these projects are described. The particular importance of the communication of risk is identified and some conclusions are provided with respect to the future for this method of procurement.

In many respects the considerations of the Geotechnical Engineer are no different to those applying to other disciplines. However, the added pressures on the Geotechnical Designer are that their work is necessarily early in the programme and that the ground is generally one of the areas of biggest risk to the project. The ground-related issues can also offer the greatest opportunities for securing the project at tender. In addition, the relationship between the Designer and Contractor may result in a very limited geotechnical presence on

Transportation geotechnics. Thomas Telford, London, 2003

site during the contract to deal with issues and check actual conditions. This requires the Designer to ensure that adequate guidance is provided within the documentation, not only to implement the design but also to deal with the unexpected.

The advent of 'Early Contractor Involvement' (ECI) may well herald positive changes for our Contractor Clients and reduce the relatively high risks faced by them and their Designers. Only time will tell if this is the case.

The PFI framework

In outline, the participants in PFI projects are as summarised in Figure 1:

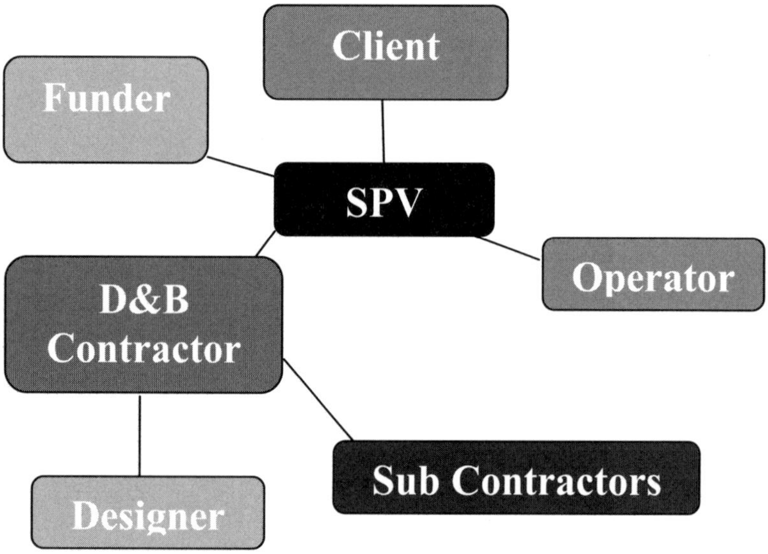

D&B Contractor – Design and Build Contractor
SPV – Special Purpose Vehicle

Figure 1, The structure of a PFI project

The priorities of each of the participants are aligned in many areas, such as completing on time, ensuring a positive cash-flow and maximising profit, and there is a general principle of risk transfer down the contractual chain wherever possible. However, there are also priorities specific to each of the participants. For example, of particular relevance to the D&B scenario is the Operator's concern with plant/ building life cycle cost as well as reliability.

The Designer is usually at the end of a long contractual chain. This means conforming to a number of different requirements imposed by other parties, with little ability to influence or change them. In addition, the Geotechnical

Team is likely to be part of a much larger team and subject to its particular requirements.

Tender input

General features

All participants will be interested in minimising bid costs and a balance must be reached between this and solution certainty or risk. The tender timescales are relatively short varying between about 14 and 26 weeks and significant tender 'addenda' can be anticipated throughout this period. These 'addenda' often include additional ground investigation information which may be critical to the geotechnical design.

Due to cost, a full design cannot be carried out at tender stage which means that the quantification is unlikely to be accurate. However, the geotechnical design input needs to be sufficiently detailed to gain confidence in the proposals. At this stage, significant "Value Engineering" may be carried out in order to try and win the contract.

Once the contract is secured, there can be very severe penalties for non delivery or for increases in construction costs. Therefore, the Contractor (the Designer's Client) has to understand the risks inherent in the tender design and be able to price them. Clearly there is a fine balance to be met, since an overestimate will result in an unsuccessful tender whilst an underestimate will mean losing money.

Tender design input is generally provided below cost and the almost ubiquitous late changes to the Employer's Requirements generally mean that even the most well organised tender design input costs more than originally proposed. Clearly this is not just a problem for Designers since the Contractors have to bear most of the design costs at risk.

Tender process

Prior to undertaking any work, the design team needs to gain a thorough understanding of the contractual obligations being entered into. The input required will of course be dependant on the size and complexity of the project. The following illustrates the typical process for a PFI project:

- Familiarisation – this may require a detailed review of a number of different contractual conditions, which may be cross-referenced in the tender documents. In addition, a site visit is carried out and a 'Desk Study' is prepared if needed;
- Brainstorming - participants at a senior level from both the Contractor's and Designer's organisations meet to try and identify all design options available for the project;

- Choice of solution(s) – advantage is taken of any particular 'specialisms' residing within the team and preferred design solutions are identified for further development;
- Tender ground investigation – depending on the level of information available, additional physical investigation is carried out either in collaboration with the other tenderers or in 'private' depending on the design options being considered;
- Preliminary design - in terms of geotechnical design this is likely to include production of geotechnical 'long sections' and 'feature plans', material classification, design of earthworks solutions, structural foundations, identification of testing and monitoring schemes and the production of a tender report;
- Identification of risks and opportunities – throughout the tender period all participants contribute to a risk and opportunities register, (discussed in the next section);
- Consultations – regular consultations are held with the 'Client' body (SPV) to review progress on the tender and clarify issues as they arise during the preliminary design;
- Departures from standards – any proposed variations to the identified specification are submitted for approval;
- Quantities – these are prepared based on the preliminary design;
- Sub-contract enquiries – usually made by the Contractor but based on specifications and principle quantities provided by the Designer. In the case of proprietary systems, design discussions are held with the specialists to ensure the suitability of the product being offered;
- Estimation – the tender Bill of Quantities is prepared usually by the Contractor;
- Costing of risk – residual risks not covered by other tender quantities are reconciled;
- Quality and financial tender submission.

Risks and opportunities

At tender stage, risk registers are used for conveying uncertainties in the preliminary design. Solutions, which reduce cost, are obviously key but those that reduce construction time are equally important. A Contractor Client will want to be involved in the choice of design solutions and must also be fully appraised of the associated risks. The risk register should be used as a key document to present the geotechnical uncertainties, both positive and negative in terms of their effects on cost, that are present in any design. The Designer will of course also have to provide tender design drawings and written advice for any solution including quantitative information. These will be based on a balance of probability and cost. The Designer then uses the risk register to make the Contractor aware of other possible scenarios and the cost and programming implications. The Contractor is then able to judge how much risk capital they will include in their tender submission.

Where significant opportunities are identified further "Value Engineering" design input may be agreed and the solutions can be honed via an iterative Designer/ Contractor input. The tender team seeks to optimise the cost, programme and buildability of the scheme in an effort to secure the project. At this time, risk workshops are often held to bring all parties together to resolve any issues. Where possible, risk is of course avoided but where this is not possible, it is either transferred or mitigated against by further design input, or by qualification or allowance within the tender itself.

No matter how experienced the Contractor, it is important to identify any factors that could affect project time and cost and consider these in the risk register. Aspects that might seem obvious; for example, allowing for the impact of earthworks on structures in the construction programme or for the monitoring of ground movements, these need to be detailed in the tender. The typical issues considered by the risk register are:

- All design risks (either positive or negative);
- the likelihood of the issue arising
- the consequence of this – time or cost;
- the control measures to be adopted;
- the risk following control;
- the risk owner;
- consideration of any residual risk;
- cost.

Whatever the methodology, the risk will have to be evaluated and an assessment made of the provision to be allowed within the tender. To facilitate this, there clearly needs to be a consistent approach to risk assessment across all the disciplines within the project team.

Some Contractors, particularly in larger projects, will use complex models such as 'Monte Carlo' simulations to evaluate the project risks, whereas others will take a view based on experience. Whatever the Contractor's approach, it is important that a dialogue is entered into prior to tender submission so that there are no misunderstandings regarding the meaning of the risk assessments made.

The Designer is usually offered no alternative than to use a risk register, since providing a 'no risk' proposal will inevitably lead to an unsuccessful tender submission.

As well as the direct risk to the tender design itself there are often significant commercial risks for the Designer organisation associated with the nature of PFI contracts. In such circumstances, the only appropriate response is a properly evaluated assessment of risk to the design organisation which can then be discussed with the Contractor. The risk can then either be transferred or built into the design fee offered.

Assuming that the tender is successful, the Designer will need to ensure that the risk register is considered as being part of the advice provided at tender stage. The Contractor may otherwise assume that all the design is covered by the tender drawings.

The design process

Geotechnical design criteria for PFI and D&B projects are likely to include the following features:

- Fast track;
- insufficient ground investigation (GI) – hence creating a need to make assumptions and confirm them later, with additional GI which starts as detailed design starts;
- potential for 'optioneering' by the Contractor;
- many stakeholders;
- varying degrees of Contractor involvement in the design process;
- slow approval processes (whilst Contractor and ultimate Client consider options);
- third party certification;
- designs being subject to limited Designer verification on site.

Additional ground investigation

The 'fast track' nature of D&B projects means that geotechnical design invariably starts ahead of receipt of the full ground investigation data. Often additional ground investigation work is designed and undertaken in tandem with the preparation of a geotechnical interpretative report as well as the geotechnical design. The Designer is therefore forced to consider very carefully the risk and delays associated with undertaking additional GI, versus those of proceeding without it. The risks of undertaking additional GI include:

- Delays in obtaining data - the procurement processes may result in a 6 week period before the additional GI commences, so a period of 10 weeks minimum should be allowed until the site and laboratory data is available for design. It must be considered whether the programme can accommodate such a delay;
- The data may invalidate the proposed tender design – this will not be well received by the Contractor and Client after the delay in obtaining the information;
- Requirement to provide certification - often a geotechnical interpretive report is required to support the additional GI which may cause further delay to design and thus cost implications.

The risks of proceeding with just the tender GI data include:

- The design may not be the most economic (but this is often outweighed by the programme issues);
- The design may not be technically acceptable unless proven by additional GI.

The balance of these risks and particularly the time aspects may often be such that additional GI is limited to the use of techniques that can be completed relatively quickly (such as Cone Penetrometer Testing) or to isolated locations such as structures, where the available data is inadequate. These data can then be used to justify the design improvements offered at tender stage. Wherever possible, the Designer is likely to seek to include this GI as part of the works, to minimise certification requirements due to the adverse effects that such works can cause to a programme. This approach to additional GI may not always be the best solution for all parties, but is a consequence of the contractual arrangement (Figure 2).

Figure 2, The Designer's dilemma

The AGS (Association of Geotechnical Specialists) database format

The handling of geotechnical data from many sources, often in great quantities, requires careful management. The use of the AGS format is a key tool in the Geotechnical Designer's armoury in this respect. Indeed in the Authors recent experience, the use of the AGS format was the only way that the vast amount of data could be handled to meet the programme. In some cases this has involved converting historical borehole and test data to allow the preparation of a common tool for the project.

Design approach

The Designer must always remember that the fundamental requirements are to produce documents for construction and not get overly involved with the 'design' as an end in itself. As with other elements of design, the geotechnical constraints require that the designs should be simple and buildable with maximum reuse of site-won materials. The focus here is on programme, which usually means that a 'moderately cautious' design approach is appropriate. As a result, the technical effort is often concentrated on selected critical elements with 'standard designs' being used elsewhere as far as possible. Consideration must also be given to the time required for checking and approval since the more inventive or unusual the design is, the longer this process is likely to take. Indeed if 'departures from standards' are identified at this design stage, very long lead in times can be anticipated before an 'approval in principle' is forthcoming.

Where detailed designs differ from tender design it is vital that these changes are justified and agreed with the team. Again, because of the contractual framework, there is a risk that aspects that should have been picked up at tender stage may result in a claim from the Contractor.

Once the construction starts on site, the Contractors time-related costs or penalties will generally dominate all but the most significant design-generated savings. Innovations in design, if not included in the tender design, may not be welcomed at this stage unless they are of significant time and cost benefits. Any cost savings that may be identified during the detailed design are usually subservient to the time implications that may arise in their construction.

The output in terms of specifications and construction drawings often means that the Geotechnical Engineer's designs are incorporated into documents prepared and issued by others. Generally, it is preferable for the Geotechnical Designer to have clearly defined deliverables directly under his control. This can sometimes extend to being directly responsible for all related outputs such as the earthworks drawings which, if provided by others, can often take a lesser priority under the pressure for delivery that occurs in D&B.

Sub – contract designs

Even smaller D&B contracts often include an element of specialist/ sub contract design. Usually the contractual framework means that the Designer adopts sub-contract designs and therefore they need to check such input. It is important that early attention is paid to the preparation of sub-contractors' design specifications and to the understanding and checking of their designs. This is vital in geotechnical design, where the input parameters can be subject to debate, particularly when a full Geotechnical Interpretive Report (GIR) is not available. In such circumstances it is usually better to agree the basic soil and groundwater design parameters (via an 'Approval in Principle' – AIP) and also require the specialist to consider not only internal stability but also global stability issues. This avoids potential conflicts between these two elements which are sometimes artificially separated.

Planning the design

Given all the pressures and constraints on the Geotechnical Designer, it is vital that the input provided within the design process is adequately planned and monitored. This usually requires detailed planning of the design process to ensure that not only are the technical issues addressed but that the interfaces with the remainder of the design team are properly understood and identified and the implications of this on the construction programme are considered. The cross disciplinary design interfaces, such as structures or drainage design, demand proper integration with the design teams working in parallel. Key issues need to be resolved with these teams at an early stage to avoid constant changes or an overly iterative design process.

Document control and quality assurance (QA) is a key aspect of the design process which will involve reviewing and checking procedures. The general approach should be to keep it simple. Complex QA tends not to work, or rapidly falls into disuse by the project participants. The use of Project Extranets are becoming more common and can be invaluable in 'bringing together' the disparate resources that are involved in the larger D&B projects. However, speed of setting up and down-loading of information, particularly drawings, can be an issue.

Teamwork

Success at detailed design stage relies on good teamwork within the D&B team. This may be assisted, particularly on larger projects by co-location of the various participants, at least until relationships between the team members have been established. The Geotechnical Designer must engage with all other relevant members of the team and avoid the criticism, sometimes levelled at specialists, of working in isolation of others.

A review of the design options and consideration of buildability issues should be discussed with the construction team at commencement of the design. However, the main decision makers are unlikely to have been appointed at this stage of design, so limited input may be given. Therefore, one of the main key benefits of D&B contacts, (integration of design with the construction team), can sometimes be severely diluted. Where such advice is given, it can be counterproductive unless the Contractor's Design Liaison Team is fully integrated with the Construction Team. Where this is not the case, problems can arise later when the site team review the designs and require late changes to be incorporated. Therefore, communication between the Designers and the construction team should be facilitated where at all possible, which is sometimes overlooked in a design environment subject to programme pressures.

The Contractor will normally be interested in as much 'optioneering' as possible at detailed design stage, to try and reduce the construction costs. Therefore, as at tender stage, careful management of this process is required to ensure that resources are not diverted away from the main task such as issuing construction deliverables. It may well be that the scope of such 'optioneering' has to be limited, based on the design fee and assumptions made at tender stage.

A key member of the team at this stage from the Designers view, is the Contractor's Design Liaison Engineer who interfaces with his construction design team, whilst offering a degree of protection from the innumerable queries and requests that arise as soon as the site is mobilised. At this stage the Designer must be allowed time to complete the design without constant requests to investigate new options identified by the newly appointed construction team. If options worthy of further investigation are identified, then to avoid delays to the main work, (i.e. delivering the design for certification and construction), this work should be carried out by a parallel design team.

It is also important that an understanding of the checkers/ relevant authorities' requirements and priorities is established to avoid misunderstandings at a later date. Some authorities will allow design solutions to be submitted as drafts and are prepared to make comments at that stage such that, once the formal submission is made, acceptance should become a formality.

Construction

It is self evident that good liaison and checking is required to ensure that the design is appropriately conveyed to the site team. A great deal of foresight is needed here since it must be recognised that the level of design presence on site is likely to be limited. Consequently little opportunity exists to identify changes from assumed ground conditions or to correct obvious discrepancies in the design. It is therefore vital that all assumptions are recorded on the construction drawings. The information to be recorded should include assumed ground conditions, assumed construction sequence, testing methods and procedures to verify assumed ground conditions. Guidance should also be given with respect to reporting requirements and protocols for dealing with unexpected conditions such as dealing with groundwater, removal and replacement of soft spots or ultimately, when reference back to the Designer is required. This is important, as frequently these drawings will be the only information the site construction team will see.

The Geotechnical Designer should make the strongest representations to ensure that a Geotechnical Engineer is included as part of the Designer's Site Representative team, particularly during the early works. This is critical to allow designs to be issued with certain items to be finalised once the earthworks are underway; for example, slope drainage requirements or the final pavement sub-grade assessment should reflect the findings on site since there will never be time to resolve all issues prior to construction unless the design is overly conservative.

The pressure on the construction team to reduce costs whilst maintaining buildability often means that the Designer's site team will be inundated with requests for design changes. In order to respond to these and maintain a verifiable design, the Designer's site team must be in possession of the full facts relating to the original design. In this regard, the earthworks and structure summaries (EAFs and GSFs etc.) provide the starting point to the geotechnical assumptions and constraints pertaining. Where significant design changes are being requested, it is far better that these are referred back to the Designers who are far better equipped to deal with detailed issues. For less significant issues, the Contractors staff (especially the materials engineer) must work as a team with the Designer's Site Representative if there is to be any scope to improve upon the design whilst on site.

The environment on site is one where the Designer has very little control. With the advent of Contractor self-certification, the Designers role is at best one of inspection to "police" the works and audit the Contractor's records (inspection of 10% of the works is a common requirement which requires a significant presence on site to achieve). This QA role has to be achieved alongside resolving technical queries and any design developments required. The level of Geotechnical Designer checking on site is therefore very limited and even then, the personnel present are often diverted towards investigating Contractor changes or clarifying the designs. This places great reliance on the Contractors self-certification scheme and the inspection, testing and monitoring plan. As previously identified, it is vital that the Designer's own detailed requirements are conveyed to the site team so that they may be properly incorporated into the Contractor's site protocols. Therefore, the Designer's assumptions and requirements must be absolutely clear in the documents issued for construction.

Since at times the construction proceeds at risk (in advance of the certified designs), changes on site can be common place and errors are sometimes made. Usually, the pressure is then on the Designer to justify what has already been built. However, the Designer must be confident of the long term performance before agreeing to an innovative design / construction approach.

Checking

Each participant to the contract will want their own representatives checking the various aspects of the work and, if not properly managed, this can lead to long delays and confusion regarding the project requirements. Fortunately, the checking process is usually streamlined with one organisation taking a lead role in co-ordinating the responses of the other parties. Even so, tried and tested

Geotechnical asset management

Dr. M. Raybould
Scott Wilson Geotechnics

Introduction

The purpose of this paper is to discuss earthworks asset management in relation to transportation corridors in the UK.

Historically, earth structures have not been considered as assets and have received a lower priority for maintenance and renewal funding compared to structural assets. However, as the nation's infrastructure network matures, greater emphasis and resources are being placed on geotechnical asset and evaluation maintenance.

This paper addresses the regulatory and commercial reasons for undertaking effective asset management, the components of a risk management approach to asset management and current techniques being utilised, trialled and researched by the major infrastructure owners.

Earthwork assets – definition and ownership

An earthwork asset (or liability) is any geotechnical structure for which an owner has responsibility. Assets are normally categorised as;

- Cuttings (rock and/or soil),
- Embankments,
- At Grade,
- Structure (e.g. retaining wall at the toe of a cut slope).

At grade sections are considered because the formation is a geotechnical asset which may deteriorate with time (e.g. settlement / collapse / liquefaction) and because any asset database should form a continuous linear record. In the UK the major infrastructure owners of Geotechnical assets are;

- Network Rail
- Highways Agency – trunk roads and motorways
- British Waterways
- Local Authorities / Councils – non trunk roads
- London Underground Limited (and subsequent Infracos)

Transportation geotechnics. Thomas Telford, London, 2003

CIRIA report 550 (CIRIA, 2001) and 591 (CIRIA, 2003) relating to condition appraisal and remedial treatment of embankments and cuttings respectively are a valuable reference source for owners and managers of earthwork assets.

Reasons for undertaking asset management

Regulations

Owners of assets have a responsibility to the public to ensure that they are maintained in a safe condition to protect employees, the public, third parties and the environment from unreasonable or unacceptable risk. This responsibility is dictated by legislation. The principal statute governing safety is the Health and Safety at Work Act (1974) enforced by the Health and Safety Executive (HSE).

The two regulations which are applicable to the maintenance of all infrastructure cuttings and embankments are The Construction (Design and Management) Regulations (1994) and The Management of Health and Safety at Work Regulations (1992).

Rail

Her Majesty's Rail Inspectorate (HMRI), which is part of the HSE, deals with all health and safety matters on the operational railway. Network Rail and London Underground Limited demonstrate their compliance with HMRI requirements by applying their own standard procedures.

Roads and waterways

The Factory Inspectorate, which is also part of the HSE, deals with health and safety outside the operational railway. Highways are regulated through the Highways Acts (1959 and 1980), which define general responsibilities of highways and local authorities with respect to highways. Waterways are regulated through The British Waterways Act (1995), Transport Act (1968) and Reservoirs Act (1975), the last applying to water stored in dams to maintain the canal network and specifically excludes the canals themselves.

Effective commercial management

In addition to important regulatory requirements, there are clear commercial benefits for asset owners by having an effective asset management programme in place.

The primary commercial benefit of asset management is to ensure assets are maintained before serviceability or ultimate limit states are reached, to avoid disruption to the operation of the infrastructure through, temporary speed restrictions on rail routes, lane closures on highways or draining sections of canal for unplanned or emergency repair.

Failure at Ultimate Limit State is easily identified as the asset can no longer be used for its intended purpose (e.g. Figure 1). Serviceability requirements primarily comprise ride quality (e.g. Figure 2), or freeboard height for canals, and are defined by the respective asset owner.

The cost of planned maintenance and repair is substantially less than responding to failure. McGinnity et al (1998) reported up to 70% savings in cost for LUL embankments and Patterson and Perry (1998) suggested cost savings of up to 80% on the M23 in Surrey by utilising a planned approach to asset management.

Approaches to asset management

No earthwork asset is risk free. Latham (1994) states that risk can be managed, minimised, shared, transferred or accepted. It cannot be ignored. A risk management strategy is a fundamental component of asset management. The main components of a Risk Management Strategy are;

- *Data acquisition* comprising inventory and evaluation of geotechnical assets.
- *Risk assessment* by qualitative and/or quantitative methods.
- *Risk reduction* through implementation of appropriate maintenance and/or engineering stabilisation measures, automated advanced warning systems (monitoring/reporting) and emergency contingency plans or further planned inspection programmes.

The ALARP principle is such that risks should be maintained 'As Low As Reasonably Practicable' and that improvements to assets should only be pursued where the cost of averting the risk is not grossly disproportionate to the risk averted. Risk assessment is a technique that should be utilised to ensure that safety objectives are met within a commercial business framework.

Data acquisition

A general risk assessment procedure (CIRIA, 2001) is illustrated in Figure 3. This procedure suggests that the data acquisition process can be divided into a strategic level and tactical level. At the strategic level all assets are identified, inspected and prioritised. At the tactical level individual assets are assessed in detail with respect to specific risks.

Strategic level

The spatial extent of assets can be determined by a variety of techniques each having limitations of effectiveness or cost;

Desk study:	OS maps
	Existing aerial photography
Rapid Inspection:	Foot
	Road / Rail (e.g. Omnicom) / Boat
	Site specific aerial photography (e.g. stereo oblique aerial photography)

The initial data acquisition should be of sufficient rigour to produce a comprehensive asset register and enable the effective planning of the tactical level inspections.

Tactical level

In advance of a specific earthworks inspection its purpose must be ascertained. This may be confirmation of asset register, rapid inspection for prioritisation of further inspection or specific detailed inspection.

What to collect:

The specific information that needs to be obtained from inspection differs for rock and soil cuttings, embankments, structures and at grade sections. Typical components which need to be considered include;

Initial desk study:	historical records, ground conditions, local terrain and rainfall information
Slope geometry:	angle, height, aspect, natural slope adjacent, retaining walls
Formation:	type, condition
Slope composition:	toe, middle, crest
Slope drainage:	face, crest, toe
Catchment drainage:	catchment size, gradient, geology, surface type
Movement indicators:	slope form, geomorphology, road/track surface, aspect of trees/fences, cracking, mass movements, history of instability
Animal activity:	scrapes, warrens, setts
Vegetation:	type, extent

Figure 4 presents a check list of stability indicators developed by the author and Network Rail specifically for the rapid inspection of earthworks.

How to collect:

For field derived data, the minimum amount of equipment required to undertake the inspections includes; digital camera, clinometer, 3 metre measuring tape and 30 metre measuring tape. On safety grounds it may be necessary to work in pairs or to have specific safety personnel available.

In order for the examiner to identify his position, hand held GPS tools are frequently being utilised. The advantages of GPS include direct reporting of OS grid reference, direct input into electronic data storage system (e.g. HAGDMS) and can, in principle be used anywhere. The disadvantages include cost, variable accuracy and potential for signal loss. However it is likely that the disadvantages will be minimised by advances in technology.

As with positioning techniques, hand held electronic data collection systems are gradually replacing traditional paper recording techniques. As the cost and user-friendliness of these tools improves they will become more prevalent.

When to collect:

Earthwork inspections are best undertaken when the vegetation cover is not too dense and after a period of rain to identify water related features. Other aspects to consider are hours of daylight, weather and temperature. Ideally, inspections should be carried out in cool dry days in October and April to satisfy the criteria above. Practically, however, inspections are undertaken throughout the year although caveats may be necessary on the reporting of certain features observed.

The frequency of inspection depends on the purpose and condition of the asset. Assets in a particularly poor condition may require annual assessment and those that, after initial strategic level inspection, appear to be in the best condition may only require revisiting every ten years. Highways Agency require that Principal Inspections are carried out on a five year cycle so that 20% of a particular managed area is completed per year.

Who should collect:

It is preferable that experienced geotechnical engineers or engineering geologists are used to undertake earthwork inspections. Network Rail and Highways Agency have developed procedures to provide less experienced staff to be given specific training to undertake inspections coupled with audit and checking by more experienced staff.

Risk assessment

Terminology

Hazard:	a component of an asset with a potential to cause an event
Probability:	the likelihood that the event results in a consequence
Consequence:	the outcome of an event
Risk:	the risk posed by a geotechnical asset (probability x consequence)

The risk associated with a hazard is a function of the probability and the consequence of the hazard occurring. A hazard may be, for example, terracing or bulging on a cutting slope due to creep movements over many years. The probability of ultimate failure may only be acceptably low as the creep has been occurring over a long period of time. The consequence of failure, (significant volumes of soil spilling on to a rail line with limited clearance), could be very high. The risk to the infrastructure (being the product of probability and consequence) could therefore be significant.

Interpretation of risk can be undertaken in a qualitative, semi-quantitative or quantitative manner. All assets should be assessed in a similar manner and prioritised in accordance with the magnitude of the risk.

Qualitative

Table 1 shows a simple qualitative (observational) approach to risk assessment where the probability and consequence of failure are determined in a subjective manner, generally comparing one asset to an adjacent one or a standardised norm.

Table 1, Example of qualitative risk assessment matrix.

		Probability of Failure		
		Low	Average	High
Consequence of Failure	Low	Negligible Risk	Low Risk	Medium Risk
	Medium	Low Risk	Medium Risk	High Risk
	High	Medium Risk	High Risk	Unacceptable Risk

The resulting risk category would determine the extent of further action, for example;

Unacceptable - remedial action with highest priority
High - remedial action within 5 yrs
Medium - remedial action may not be necessary – preventative within 5 years
Low -review value for money of preventative measures
Negligible - re-inspect in 5 years

Network Rail (RT/CE/P/030-1997 – 'Management of Embankments and Cuttings', Figure 5) and Highways Agency (HA 48/93-1993 - 'Maintenance of Highway Earthworks and Drainage'), have used similar techniques to prioritise risk, although both procedures are now superseded.

Semi quantitative

Scott Wilson developed a semi-quantitative risk assessment methodology for earthwork risk prioritisation (Figure 6) which comprised the calibration of a qualitative visual inspection and quantitative evaluation of slope failure trigger factors.

The qualitative procedure involved site inspections, by Senior Engineering Geologists/Geotechnical Engineers, collecting data as described earlier and interpreting such, in accordance with Table 1.

The quantitative procedure involved attributing numerical values to parameters that trigger slope failure, namely:

- Slope angle
- Slope height
- Slope angle adjacent to earthworks
- Catchment length
- Terrain type
- Underlying geology
- History of slope failure
- Mean annual rainfall

A comprehensive desk study of slope failure history is undertaken, to complement the field inspections. To evaluate the significance of a parameter on the distribution of slope failures, the observed number of slope failures (O) is divided by the expected number (E) based on the percentage of the total length of earthwork. Higher O/E ratios indicate a greater occurrence of slope failure than would be expected from a random distribution. The overall significance of a factor in influencing the distribution of slope failures is determined using the "Chi-square test". A Hazard Index (HI) is evaluated for each earthwork based on the ranking and weighting attributed to each parameter.

The calibration involves comparing the Hazard Index (HI) with the qualitative risk rating, resulting in three categories;

- Most Concern - High or unacceptable risk and/or high HI
- Some Concern - Medium risk and intermediate HI
- No Concern - Low or negligible risk and/or low HI

'most concern' means earthworks have a particularly adverse combination of one or a number of parameters; geometry, rainfall, catchment/run-off, geology, and history of slope failure such that they could present a risk of slope failure in the short-term.

'some concern' means earthworks have fewer adverse instability factors, but conditions are such that risk of slope failure may develop with time.

'no concern' means earthworks are unlikely to present a risk of slope failure in the foreseeable future.

The limitations of the semi-quantitative technique described is that, although the likelihood of failure is considered, the consequences to the asset of such failure is not directly addressed.

Quantitative

Quantitative Risk Assessment (QRA) follows a logical procedure based on the principles of an event tree analysis, where numerical scores are attributed to the probability and consequence of each potential hazard to cause harm resulting in a numerical score of 'risk' to the infrastructure.

Clearly, the potential hazards associated with rock cuttings, earthwork cuttings, and embankments and their consequence on the infrastructure in the event of failure are significantly different and the risk assessment approach taken for each will be different.

Network Rail procedure RT/CE/P/030 – 2002 – 'Management of embankments, cuttings and natural slopes', sets out their policy on the management of earthworks to ensure that risk to the operational railway is as low as reasonably practical. The procedure states that cyclical examinations of earthworks should be undertaken in accordance with RT/CE/S/065. Although this specification has not yet been published, certain aspects of it are currently being trialed. The examination procedure considers potential slope instability factors, such as those described earlier, and consequence factors. Pre-determined numerical scores are attributed to these factors, which are different for rock slopes, soil cuttings and embankments, and the resulting scores used to generate either the Soil Slope Hazard Index or Rock Slope Risk Appraisal Value. These parameters determine the condition status of the asset as Serviceable, Marginal or Poor;

- *Serviceable* condition means an earthwork that could not reasonably be expected to deteriorate to a condition likely to impose a risk of failure within the next 10 years.
- *Marginal* condition means an earthwork that could not reasonably be expected to deteriorate to a condition likely to impose a risk of failure within the next 5 years.
- *Poor* condition means an earthwork that could reasonably be expected to deteriorate to a condition likely to impose a risk of failure within the next 5 years.

Rock cuttings – rail rock slope risk appraisal

McMillan and Matheson (1997) describe a technique and field trials for rapid assessment of rock slopes adjacent to highways. The hazard index considers primary parameters – those that have the potential for failure and secondary parameters - the likelihood and consequence of failure.

The following parameters are determined from the rapid assessment;

- Evaluated Failure potential
- Factor of safety for failure
- Discontinuity – principal spacing, trace length and dilation
-
- Observations of potential failure
- Potential failure size and position
- on rock slope
- Rock material strength and weathering
- Ground Water

- Rock trap size and shape
- Slope profile and berms
- Road sight lines
-
- Cutting type and associated hazards
- Remedial works
- Traffic volume
-

The numerical rating from the procedure results in four categories; No Action, Review in 5 years, Detailed Inspection or Urgent Detailed Design. The resulting procedure RoSHI (Rock Slope Hazard Index) has been utilised extensively by Highways Agency in their asset management programme, (e.g. Hitch, 2002).

McMillan and Manley (2003) describe a Rail Rock Slope Risk Appraisal (RRSRA) system. The general approach is based on the logic of the TRL Rock Slope Hazard Index and Quarry Rock Slope Hazard Index (e.g. McMillan and Matheson [1997]). The RRRSA is based on rapid field surveys that record essential data on the geotechnical conditions of the rock mass, rock slope geometry, track alignment and exposure to hazard. The primary objective of the RRSRA is to provide an initial classification of rock slope condition that can be used to prioritise future action, repeat surveys and manage rock slope risks (refer to Figure 7). In order to achieve rapid data collection no physical measurements are taken and all data is based on visual assessment only, resulting in each survey taking between 10 and 30 minutes. The data, collected from field investigations, are used to calculate the RRSRA value and determine the condition status of the rock cutting status as; Serviceable, Marginal or Poor.

Soil slopes – soil slope hazard index

Manley and Harding (2003) describe a quantitative risk assessment procedure for use in the rail environment. The Soil Slope Risk Value (SSRV) is determined as a product of the Soil Slope Hazard Index (SSHI) and the Consequence Index (CI).

The CI is based on the importance of the rail line, presence of rail furniture, train frequency and other network considerations.

The SSHI is established by desk study and site observations of stability indicators. There are 29 parameters with some 164 alternatives for embankments and 30 parameters with 177 alternatives for cuttings. An example

Debris flow *(Figure 12)*	1. Excavate/replace failed material with rock fill 2. Erosion protection (vegetation) 3. Improve crest and slope face drainage 4. Provide catchment toe wall
Mining subsidence *(Figure 13)*	1. Inspection/ground investigation / monitoring 2. Grouting of voids in worst areas
Rock Falls *(Figure 14)*	1. Trim/remove unstable trees 2. Scaling of rock face 3. Rock bolts/dowels to stabilise large blocks 4. Netting/mesh to stabilise ravelling/small blocks 5. Dentition 6. Rock trap fence/ditch
Animal Infestation *(Figure 15)*	1. Capture and relocation 2. Grouting of voids/burrows 3. Protection screens

Advances in asset management

Highways Agency geotechnical data management system

Highways Agency Design Manual for Roads and Bridges (HD 41/03) – Maintenance of Highway Geotechnical Assets describes the Highway Agency's procedure to assess the condition of highway geotechnical assets and the planning of a programme of repair and preventative works.The key management system for the HA geotechnical assets is the HA Geotechnical Data Management System (HAGDMS) which is a Geographical Information System (GIS) based on a mapping interface and linked data bases.

The HAGDMS system provides;

- Mapping interface, (at scales up to 1:1250 urban area and 1:2500 in rural areas).
- Basic geological mapping on a scale of 1:625,000.
- Databased index of all highways related Geotechnical reports and boreholes held by the HA and their Managing Agents.
- Databased index of boreholes held by the British Geological Survey.
- A database facility for all Geotechnical Asset inspection and maintenance records.
- Facilities for archiving copies of all geotechnical reports.

- Evaluated Failure potential
- Factor of safety for failure
- Discontinuity – principal spacing, trace length and dilation
-
- Observations of potential failure
- Potential failure size and position
- on rock slope
- Rock material strength and weathering
- Ground Water

- Rock trap size and shape
- Slope profile and berms
- Road sight lines
-
- Cutting type and associated hazards
- Remedial works
- Traffic volume
-

The numerical rating from the procedure results in four categories; No Action, Review in 5 years, Detailed Inspection or Urgent Detailed Design. The resulting procedure RoSHI (Rock Slope Hazard Index) has been utilised extensively by Highways Agency in their asset management programme, (e.g. Hitch, 2002).

McMillan and Manley (2003) describe a Rail Rock Slope Risk Appraisal (RRSRA) system. The general approach is based on the logic of the TRL Rock Slope Hazard Index and Quarry Rock Slope Hazard Index (e.g. McMillan and Matheson [1997]). The RRRSA is based on rapid field surveys that record essential data on the geotechnical conditions of the rock mass, rock slope geometry, track alignment and exposure to hazard. The primary objective of the RRSRA is to provide an initial classification of rock slope condition that can be used to prioritise future action, repeat surveys and manage rock slope risks (refer to Figure 7). In order to achieve rapid data collection no physical measurements are taken and all data is based on visual assessment only, resulting in each survey taking between 10 and 30 minutes. The data, collected from field investigations, are used to calculate the RRSRA value and determine the condition status of the rock cutting status as; Serviceable, Marginal or Poor.

Soil slopes – soil slope hazard index

Manley and Harding (2003) describe a quantitative risk assessment procedure for use in the rail environment. The Soil Slope Risk Value (SSRV) is determined as a product of the Soil Slope Hazard Index (SSHI) and the Consequence Index (CI).

The CI is based on the importance of the rail line, presence of rail furniture, train frequency and other network considerations.

The SSHI is established by desk study and site observations of stability indicators. There are 29 parameters with some 164 alternatives for embankments and 30 parameters with 177 alternatives for cuttings. An example of a data sheet is shown in Figure 8. The following primary features have been adopted to derive the SSHI;

- *Earthworks Factor*- features defining the shape and size of the earthwork
- *Actual Failure Indicators*- features which indicate the presence and severity of ongoing failure for any of the 5 failure modes.
- *Potential Failure Indicators*-factors which indicate the potential for failure of each of 5 failure modes (deep rotational, shallow translational, earth-flow, washout and burrowing).

The SSHI combines scores from the Earthworks Factor, Actual Failure Factors and Potential Failure Factors resulting in the earthwork condition being categorised as Poor, Marginal or Serviceable.

Risk reduction

On completion of the asset inspection and evaluation programme a prioritised list of features is determined based on the risk to the infrastructure resulting in three main courses of action; planned re-inspection, monitoring/reporting and emergency contingency plans or maintenance/repair.

Re-inspection

A programme of re-inspections at various frequencies should be undertaken dependent on the current condition and most probable trend, typically in the range 1 to 10 years. Network Rail (RT/CE/P/030 – 2002) suggests 1, 5 and 10 year intervals for earthworks in poor, marginal and serviceable conditions respectively.

Monitoring/reporting

Network Rail North West Zone have implemented a procedure referred to as Site Investigation, Instrumentation and Monitoring scheme (SIIMs) – (European Foundations, [Spring 2002]) to cost effectively manage the risk of earthworks that do not require immediate attention but may deteriorate in the medium term.

Site Investigation is undertaken utilising, where practical, a standard array of investigation and monitoring techniques to enable the hazard identified in the inspection to be observed over a period of time. The type and location of the techniques adopted are chosen to minimise disruption to the asset and provide safe access for personnel during installation and monitoring. A typical array of instruments for a cutting and embankment are shown on Figure 9. There are obviously a number of different SIIMs schemes which could be considered. However, to maximise the output and rewards from SIIMs, the right instruments must be installed at the right locations and the appropriate monitoring frequency established.

Monitoring of the instrumentation and the track level/alignment are undertaken at regular intervals (daily to three monthly) and deterioration reported, with emergency contingency plans in place if unexpected movements are recorded, and recommendations for future monitoring made. At the end of the monitoring programme, ideally at least one year, a geotechnical stability assessment report is produced making recommendations for either remediation of the observed deteriorating hazard, continued monitoring, or cessation of monitoring.

SIIMs has proven very successful and has enabled the North West Zone to;

- establish which sites are deteriorating
- the depth and mode of failure
- the rate of deterioration
- identify the appropriate intervention timescale
- identify, design and implement the most cost effective remedial works
- implement stabilisation schemes prior to deterioration impacting on the operational railway

Maintenance/stabilisation measures

Familiarity with earthwork and rock slope inspections and subsequent detailed assessments and design of remediation schemes has resulted in common problems and solutions being identified. Table 2 presents a list of typical hazards and remediation techniques.

Table 2, Typical hazards and remediation techniques

Hazard	Remediation Technique
Shallow slides/creep *(Figure 10)*	1. Remove failed material and/or regrade slope to a flatter gradient 2. Erosion protection/ stabilisation (vegetation) 3. Improve slope crest drainage 4. Improve slope face drainage 5. Provide toe or catchment wall
Deep-seated slides *(Figure 11)*	6. Excavate/ replace failed material with rockfill (smaller slides) 7. Install toe gabion or masonry wall and regrade slope (larger slides) 8. Anchored reinforced retaining wall (very large slides) 9. Improve crest and slope face drainage 10. Erosion protection (vegetation)

Debris flow *(Figure 12)*	1. Excavate/replace failed material with rock fill 2. Erosion protection (vegetation) 3. Improve crest and slope face drainage 4. Provide catchment toe wall
Mining subsidence *(Figure 13)*	1. Inspection/ground investigation / monitoring 2. Grouting of voids in worst areas
Rock Falls *(Figure 14)*	1. Trim/remove unstable trees 2. Scaling of rock face 3. Rock bolts/dowels to stabilise large blocks 4. Netting/mesh to stabilise ravelling/small blocks 5. Dentition 6. Rock trap fence/ditch
Animal Infestation *(Figure 15)*	1. Capture and relocation 2. Grouting of voids/burrows 3. Protection screens

Advances in asset management

Highways Agency geotechnical data management system

Highways Agency Design Manual for Roads and Bridges (HD 41/03) – Maintenance of Highway Geotechnical Assets describes the Highway Agency's procedure to assess the condition of highway geotechnical assets and the planning of a programme of repair and preventative works.The key management system for the HA geotechnical assets is the HA Geotechnical Data Management System (HAGDMS) which is a Geographical Information System (GIS) based on a mapping interface and linked data bases.

The HAGDMS system provides;

- Mapping interface, (at scales up to 1:1250 urban area and 1:2500 in rural areas).
- Basic geological mapping on a scale of 1:625,000.
- Databased index of all highways related Geotechnical reports and boreholes held by the HA and their Managing Agents.
- Databased index of boreholes held by the British Geological Survey.
- A database facility for all Geotechnical Asset inspection and maintenance records.
- Facilities for archiving copies of all geotechnical reports.

The basis of the risk assessment procedure remains a qualitative assessment of the assets' current condition and how the feature may deteriorate over the next 5 years. The system has a pro-active calculation procedure which uses the field observations of defects, instability indicators and areas of previous repair to trigger, 'at-risk', other earthworks on the network with similar geology and more severe morphology.

Remote sensing

Duffell et al (2002 and 2003) describe a literature review and trials undertaken for Highways Agency to institute a pro-active earthwork management strategy using remote sensing techniques.

The brief was to identify a technique that would most successfully satisfy two aspects; firstly, earthworks inventory compilation (location, slope height / angle etc.) and secondly, to provide data for the assessment of earthwork condition. The techniques considered in detail included;

- Vertical high resolution satellite photography
- Vertical aerial photography
- Stereo oblique aerial photography
- Helicopter and fixed wing airborne laser scanning (LiDAR – Light Detection and Ranging)
- Digital Aerial videography
- Thermal Line scanning
- Satellite and airborne radar interferometry (inSAR)

LiDAR was identified as being one of the most suitable techniques and a 20km trial on the M25 motorway was carried out. Following the trial it reported that the technique can be used to rapidly obtain data to identify slope instability features and that the most significant benefit may come from its ability to provide rapid topographic data for road improvement schemes.

Conclusion

The case for effective asset management has been made with respect to regulatory requirements and economic benefits.

Asset owners should develop a clear Risk Management Strategy (RMS) for their assets. The RMS should allow for the compilation of a strategic level comprehensive asset register, a technique for risk assessment (qualitative, semi-quantitative, or quantitative) and a procedure for managing the assessed level of risk.

Geotechnical asset management is an evolving discipline and asset owners are continually reviewing their procedures and trialling new techniques as advances in technology facilitate more rapid and cost effective methods of data acquisition and evaluation. Geotechnical engineers and engineering geologists face the challenge of ensuing that these techniques are harnessed to optimise the risk management of infrastructure earthworks.

Acknowledgements

The author gratefully acknowledges Chris Duffell, Highways Agency and Gerard Manley, Network Rail, for granting permission to publish photographs and use material obtained from their projects.

References

CIRIA. *Infrastructure embankments* – condition appraisal and remedial treatment, CIRIA C550. London, 2001.

CIRIA. *Infrastructure cuttings* – condition appraisal and remedial treatment, CIRIA C591. London, 2003.

Duffell C, Willis D and Rudrum M. *Remote Sensing Techniques For Asset Assessment,* Railway Engineering 5th International Conference. London, July 2002.

Duffell C, Willis D and Rudrum M. *Earthworks Asset Assessment Using Remote Sensing*, Railway Engineering 6th International Conference. London, May 2003.

European Foundations. *The Fleming Award – Excellent Adventures*, pp. 26-27. Spring 2002.

Hitch CJ. Caution Rockfall! – *An Assessment of Rockfall Hazards in the Peak District.* British Geotechnical Association and Division of Civil Engineering University of Dundee, Seventh Young Geotechnical Engineer's Symposium. Dundee, 17-19th July, 2002.

Latham M. *Constructing the Team.* Final report of the Government/Industry review of procurement and contractual arrangements in the construction industry. HMSO, London, 1994.

Manley G and Harding C. *Soil Slope Hazard Index as a Tool for Earthworks Management.* Railway Engineering 6th International Conference. London, May 2003.

McGinnity BT, Fitch TR and Rankin WJ. *A systematic and cost-effective approach to inspecting, prioritising and upgrading London Underground's earth structures.* Proceedings of the seminar "Value of Geotechnics in Construction", Institution of Civil Engineers, Construction Research Communications Limited. London, pp. 309-332, 4 November 1998.

McMillan P and Manley G. Rail *Rock Slope Risk Appraisal.* Railway Engineering 6th International Conference. London, May 2003.

McMillan P and Matheson GD. A *Two Stage System for Highway Rock Slope Risk Assessment.* Int. J. Rock Mech. & Min. Sci, Vol. 34:3-4, 1997.

Patterson D and Perry J. *Geotechnical Data and Asset Management Systems for Highways.* Proc. of the Maintenance Engineers Conference, Surveyor and Municipal Journal. London, 9-11th September, 1998.

Figure 1, Example of earthwork asset at ultimate limit state

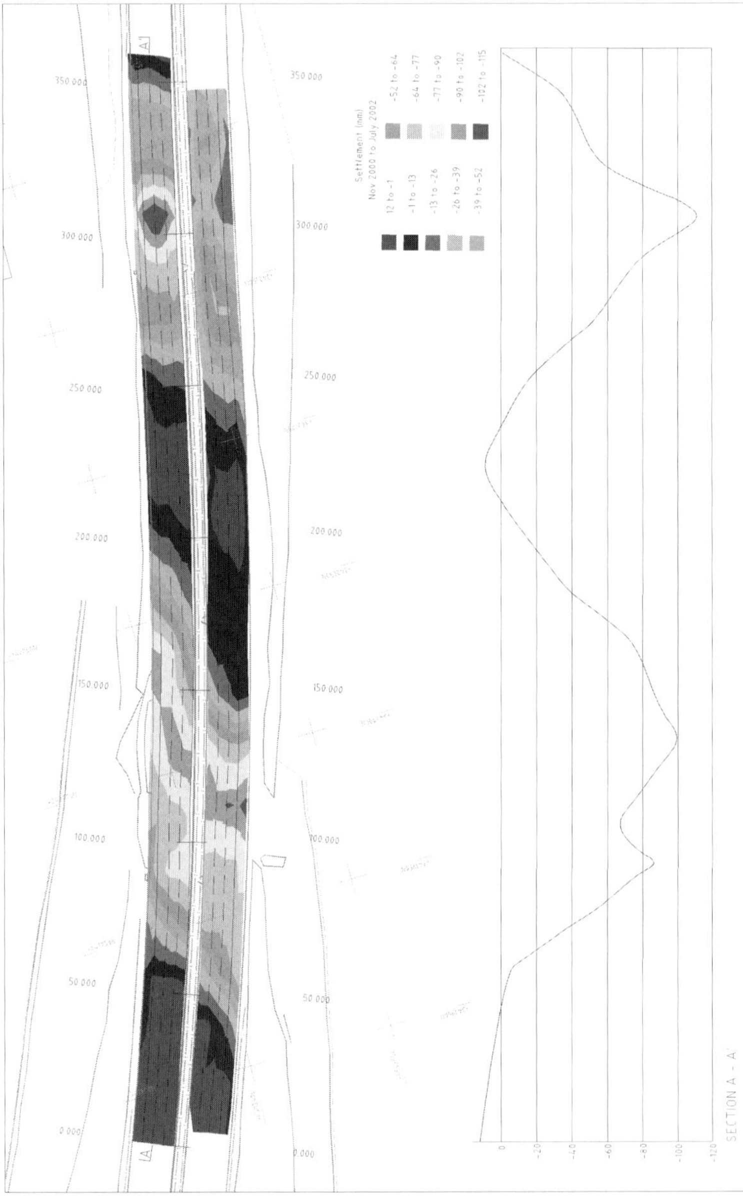

Figure 2, Example of earthwork asset at serviceability limit state

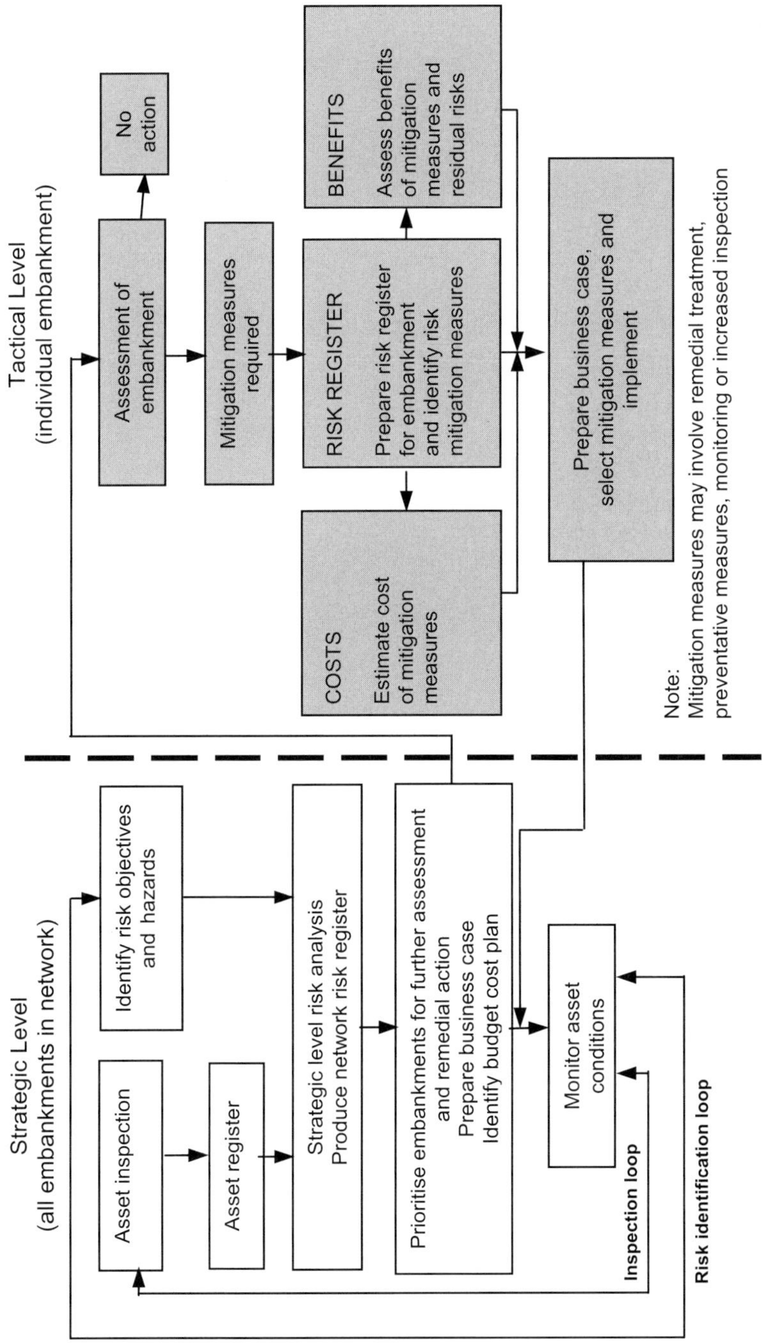

Figure 3, General risk assessment procedure (after CIRIA, 2001)

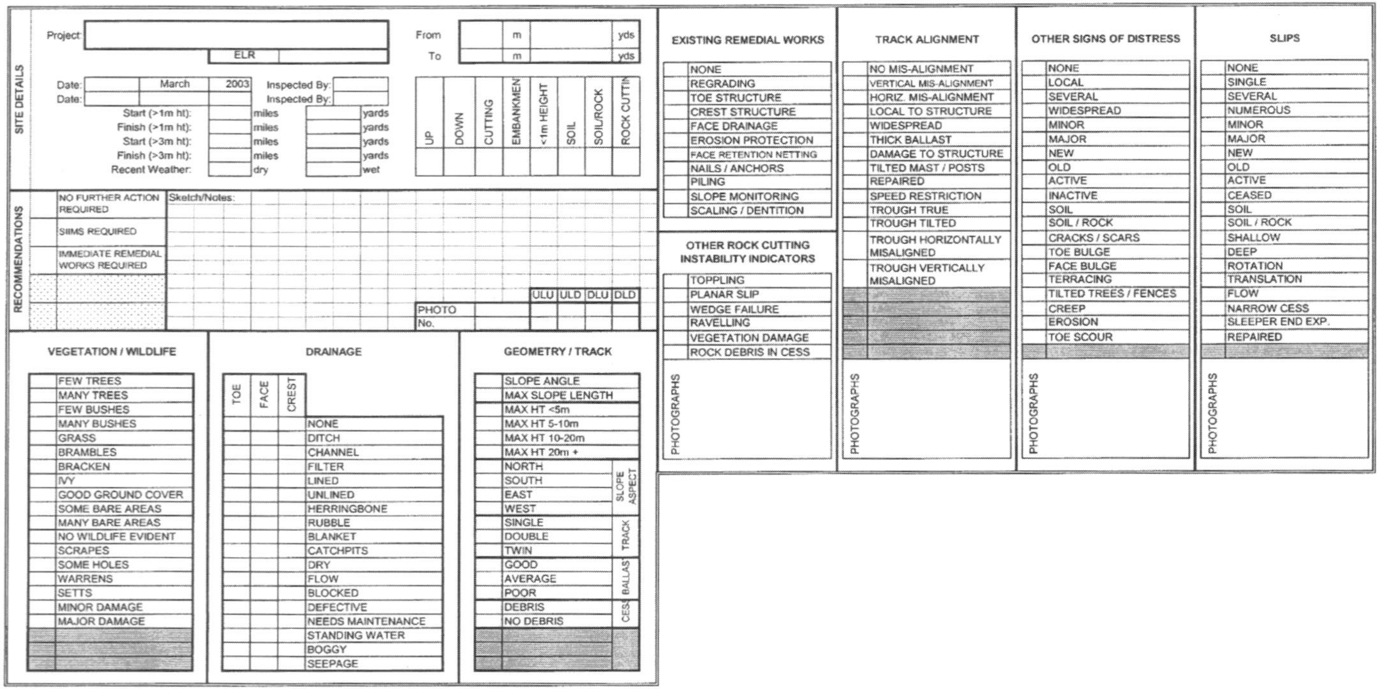

Figure 4, Stability indicators for rapid assessment of earthworks

TABLE A

Top of slope drainage	None	Functions well	Adequate	Requires maintaining	Inadequate
Bottom of slope drainage	None	Functions well	Adequate	Requires maintaining	Inadequate
Face of Slope drainage	None	Functions well	Adequate	Requires maintaining	Inadequate
Track condition		Not affected			Line and level affected
Wildlife		No evidence	Rabbit scrapes evident	Some rabbit holes evident	Warrens, setts or large holes evident
Vegetation		Vegetation giving good ground cover	Adequate veg. Cover (shrubs, trees, etc.)	Patchy veg. Cover with bare patches	Large bare patches

TABLE B

Factor	Score of 1	Score of 2	Score of 3	Score of 4	Score of 5
General condition	Expected to remain stable	Appears stable	Minor defects	Significant defects	Unfit for purpose
Trend of condition	Expected to improve	No significant change expected	Minor degradation expected	Major degradation expected	Catastrophic failure expected
Localised events affecting track	Very low risk	Low risk	Average risk	Moderate risk	High risk
General events affecting track	Very low risk	Low risk	Average risk	Moderate risk	High risk
Sum of scores for Table B					

Figure 5, Qualitative risk assessment procedure (RT/CE/P/030-1997)

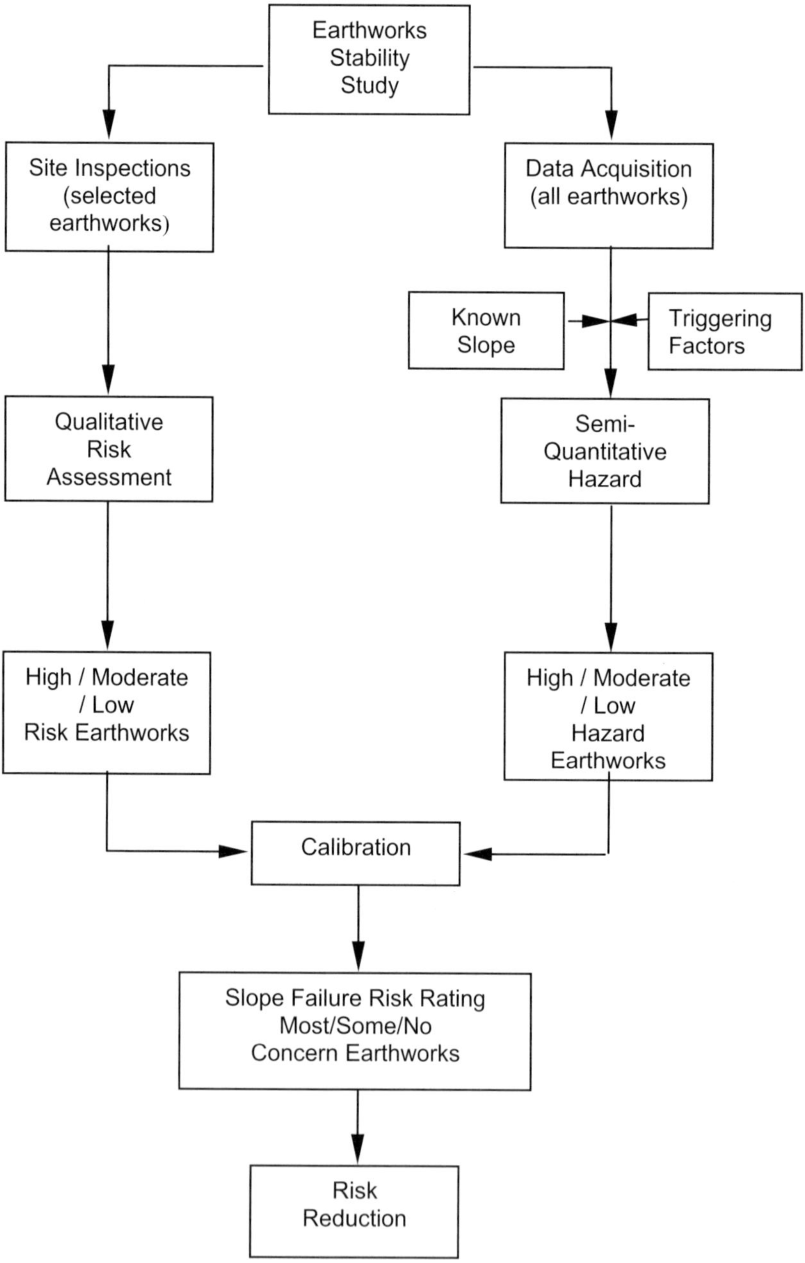

Figure 6, Semi quantitative risk assessment procedure

Data Requirements for RRSRA

Geometric

- Location
- Rock slope geometry
- Upper slope geometry
- Lower slope geometry
- Rock trap geometry

Geotechnical

- Rock material properties
- Discontinuity properties
- Upper slope materials
- Lower slope materials
- Observed potential failures
- Surface and ground water
- Vegetation

Remedial Works

- Type of works
- Coverage
- Effectiveness

Exposure

- Live rail
- Other infrastructure

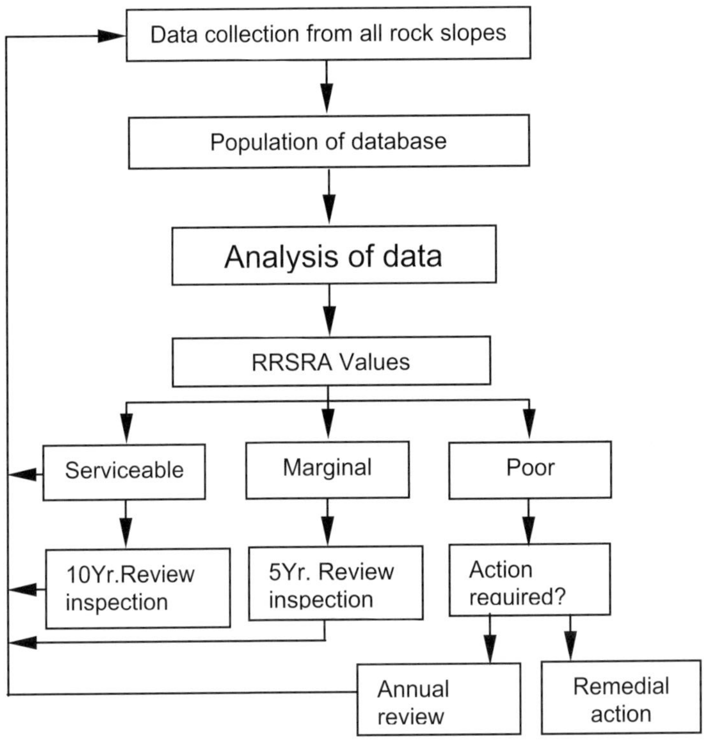

Figure 7, Quantitative risk assessment procedure - rail rock slope risk assessment index (after McMillan and Manley, 2003)

SOIL CUTTINGS SSHI FIELD DATASHEETS

STABILITY INDICATORS FOR SOIL EMBANKMENTS	STABILITY INDEX PARAMETER	OBSERVED/MEASURED VALUE	REF EMB	Parameter VALUE	SITE SCORE (add values)
Slope Geometry	Slope Angle and Slope Height	<15 degrees, <3m High	A1	0	
		15 to <25 degrees, <3m High	A2	20	
		25 to <35 degrees, <3m High	A3	40	
		>35 degrees, <3m High	A4	70	
		<15 degrees, 3m to <10m High	A5	0	
		15 to <25 degrees, 3m to <10m High	A6	20	
		25 to <35 degrees, 3m to <10m High	A7	40	
		>35 degrees, 3m to <10m High	A8	70	
		<15 degrees, >10m High	A9	0	
		15 to <25 degrees, >10m High	A10	20	
		25 to <35 degrees, >10m High	A11	40	
		>35 degrees, >10m High	A12	70	
	Slope angle adjacent to earthwork (i.e. Sidelong Ground)	(-)ve	B1	0	
		(+)ve <5 degrees	B2	5	
		(+)ve <5 to 15 degrees	B3	10	
		(+)ve >15 degrees	B4	15	
	Retaining walls 1m high or greater	None	C1	0	
		<1m height but >20m length	C2	0	
		No evidence of distress	C3	0	
		Minor distress (spalling, pointing etc)	C4	10	
		Cracking / evidence of lateral displacement	C5	15	
		Evidence of repairs	C6	15	
	Constructive activity at slope toe (i.e. excavation)	None	D1	0	
		Excavation (<10m³)	D2	15	
		Excavation (>10m³)	D3	30	
Minimum slope to track seperation	Distance between sleeper ends and crest of embankment	Embankment cess width >6m	E1	0	
		Embankment cess width 3-6m	E2	0	
		Embankment cess width 1-3m	E3	0	
		Embankment cess width <1m	E4	0	
Adjacent Geology	BGS Geological Strata Shown within 0.2km of Earthwork limits (if more than one of these is present, use the cumulative score)	Drift			
		Boulder Clay	F1	5	
		Blown Sand	F2	0	
		Alluvium	F3	10	
		Terrace Deposits	F4	0	
		Sand and Gravel	F5	0	
		Peat	F6	10	
		Head Deposits	F7	15	
		Laminated Clay	F8	15	
		Landslip	F9	15	
		Made Ground	F10	20	
		Solid			
		Manchester Marl	F11	5	
		Mercia Mudstone	F12		
		Shale/Mudstone Carboniferous	F13		
		Other Competent	F20	0	

Figure 8, Quantitative risk assessment procedure – soil slope hazard index (after Manley and Harding, 2003)

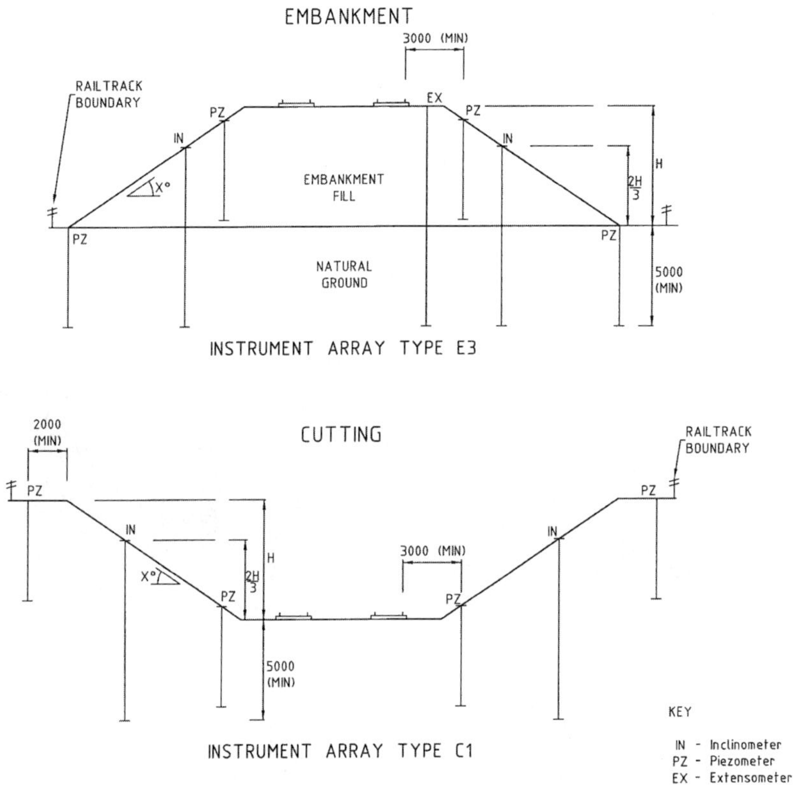

Figure 9, Typical instrumentation arrays for site investigation, instrumentation and monitoring scheme

Figure 10, Shallow / creep failure and repair

Figure 11, Deep seated failure and repair

Figure 12, Debris flow failure and repair

Figure 13, Mining subsidence and repair

Figure 14, Rock fall failure and repair

Figure 15, Animal damage and repair

A performance specification for pavement foundations

Dr. P. R. Fleming
Loughborough University
Professor C. D. F. Rogers
The University of Birmingham
Dr. N. H. Thom
Scott Wilson Pavement Engineering
Dr. M. W. Frost
The Nottingham Trent University

Introduction

Taxation of virgin aggregates and on disposal of wastes in landfill is increasing the pressure to use novel, marginal and recycled materials in earthworks and "engineered fills" which form the foundations to our transport infrastructure. Traditionally, an empirical "recipe" specification approach is used for the design and construction of these foundations. However, this approach, which closely controls allowable materials, does not consider the appropriate performance properties of these materials and restricts the use of marginal or recycled products. In addition, it does not fully exploit the properties of good quality or stabilised materials that may be used. Therefore, the UK Highways Agency is working towards the introduction of a performance specification for pavement foundations to replace the current empirical approach. This paper details the performance required of pavement foundations and materials, the philosophy of the performance specification that has been developed and addresses the issues associated with its implementation. It concludes that a two-stage implementation of the specification should be adopted and that the specification could be extended to any application that requires construction of a platform/foundation to support a pavement, such as industrial paving, railway track-bed or general fill.

Transportation geotechnics. Thomas Telford, London, 2003

The functions of pavement foundations

For pavement foundations to perform adequately they must fulfil the following three functions:

- They must support a limited number of vehicles during the construction of overlying layers. The foundation must not deform excessively (i.e. undergo resilient deformation) under trafficking so as to reduce the effectiveness of the compacted structure and they must attenuate the high stresses applied by the wheels of construction traffic (directly on the foundation) to a sufficiently low level to ensure that the subgrade does not sustain significant permanent deformation (rutting).

- They must provide an adequate base for the placing and compaction of the overlying layers (i.e. they must not undergo large resilient deformations under the action of the very high compaction stresses such that the effectiveness of the compacted structure of the overlying layer(s) is reduced).

- They must provide adequate support to the overlying structural layers when the pavement is in-service, and distribute the very large number of (lower) stresses pulses transmitted through these layers to reduce the applied stress to the subgrade to a sufficiently low level. If they do not, flexural fatigue cracking of the upper layers can develop and propagate and the progressive accumulation of permanent strain (rutting) in the subgrade after very many small stress applications may lead to deterioration of the entire pavement.

In addition, the materials used must possess chemical and physical stability in the long-term and the overlying materials must provide frost resistance to the subgrade. Conditions one and two above are short-term construction conditions featuring relatively few passes of high magnitude applied stresses. Condition three is a long-term design requirement to resist many millions of lower magnitude stress applications, although it has to be recognised that the foundation materials will be in a different condition in the longer term.

To perform satisfactorily at all stages in its life the pavement foundations must possess the two primary performance parameters of adequate stiffness (or more accurately resilient elastic modulus) and resistance to permanent deformation.

Current specifications

The current UK specification for road foundations (MCHW, 1998) is based on a recipe approach, whereby selected materials (capping and sub-base) are laid and compacted with specified plant in a specified manner to achieve a minimum level of performance. The material selection is based on material grading and durability requirements, while its compaction is controlled by using a method specification detailing specified plant layer thickness and operation. This specification, particularly the series 600 section, forms the basis of material selection and placement for the majority of earthworks materials and support layers in construction across a wide range of infrastructure projects (for example, railways, "local" roads, heavy duty paving and industrial flooring). Within this approach the performance parameters defined above are not *measured*, but are *assumed* to have been achieved from previous experience.

In this specification pavement foundation design thickness is based primarily on the use of the California Bearing Ratio (CBR) to characterise the subgrade, capping and sub-base materials. Therefore, CBR is used as a measure of both material strength and stiffness. A level of performance is then assumed for the layers constructed in accordance with the specification (15% CBR for capping and 30% for sub-base), these values being then used as the basis of design for the structural layers above. These overlying layers are increasingly being designed using analytical techniques, but the full benefit of this is limited by the assumption of the 'as built' properties of the foundation.

Although the use of CBR as a performance parameter is widely acknowledged as being not wholly satisfactory (Brown, 1996), CBR has been correlated with pavement performance over many years, in many countries, and has arguably provided a trusted empirical indicator of adequate material behaviour. However, this empirical approach is clearly limited by the nature of the ingredients that will make up the recipe to give appropriate performance.

More recently, "end product" specifications have been introduced for applications such as the construction of landfill liners and structural fills. These rely on inferring properties from other measures, such as water content and density. In the case of a landfill liner the control of the compacted state ensures that appropriate permeability and strength has been achieved within the liner. This is an appropriate strategy where strong correlation exists with the required parameters. However, in the case of pavements, which are subject to complex dynamic loads and layer interactions, the required parameters (resilient stiffness and permanent deformation) cannot be accurately inferred from other simple measures and must therefore be measured directly.

The philosophy behind a performance specification for pavement foundations

A performance specification aims to provide a client with an assurance that what is being paid for is being provided. This is preferable to an indirect empirical measurement based upon experience, since this will only allow one to assume that the product *should be* acceptable. A performance specification can only be produced if there is a means of quantifying, by direct measurement, the performance of the as-constructed product. If this is possible, the client can then specify simply that the product, and the materials from which it is made, meet certain measurable criteria. This gives the manufacturer of the product freedom in both how it is made and what it is made from, which in turn creates opportunities for innovation and/or savings (eg. in materials, process, or time). The production process may thereby be made more efficient and economic. In the case of pavement foundation construction, there are additional environmental benefits to be gained by widening the range of possible materials or by enabling the full use of the potential properties of the foundation, which will allow savings to be made in the construction of the overlying structural layers.

The design of the product, i.e. the pavement foundation, requires target values of the performance parameters to be defined. The target values (and hence design requirements) are different for the short-term (construction) condition than the long-term (in-service) condition. This is a result of the different loading and environmental conditions (discussed later on). The specification is wholly dependent upon having accurate methods of performance measurement for samples of subgrade and the overlying materials in the laboratory for design purposes, as well as parallel methods for measuring the performance of undisturbed and as-laid materials *in situ*.

Therefore, to assess such performance within a performance-based specification, the following must be available:

- a means of measuring the performance parameters of the subgrade in the laboratory for both the short-term (construction) condition and the long-term (in-service) condition,
- a method of accurately predicting environmental (water content) changes in the pavement over the long-term,
- a means of incorporating the measured parameters in the design process, and
- an ability to measure the *same* parameters for the subgrade and pavement foundation layers in the field, in order to assess compliance with the design, and to facilitate the setting of suitable target values for construction which will provide assurance of the quality of the final product.

This in turn requires an understanding and assessment of both the resilient elastic modulus and resistance to permanent deformation of the materials individually and as a composite for both the construction and in-service conditions.

Pavement foundation loading and material performance

The stress pulse generated when a vehicle wheel travels across a pavement consists of vertical and horizontal stress components with an approximately sinusoidal (double) pulse of shear stress (Brown, 1996). This stress pattern subjects an element within the pavement to a rotation of principal stresses. This pulse varies with the speed, load and direction of the vehicle, and becomes repetitive with the passage of more wheels. The foundation design included in the UK specification (MCHW 1998, based on Powell et al, 1984) assumes for construction trafficking 1000 passes of an 80kN standard axle. It should also be noted that, generally, the lower the element of interest lies in a pavement, the lower the applied stress (due to the load distribution capabilities of the layered structure).

Performance of pavement materials under repeated loading

When pavement materials are subjected to a cycle of stress they sustain deformations (i.e. strains), which consist of resilient and permanent components. The magnitude of each of these deformations defines the performance parameters of the materials. The resilient elastic modulus is calculated from the resilient strain and the change in stress, usually measured on unloading. The permanent strain is found from the permanent change in dimension (i.e. the length of an element), and progressively increases under repeated stress cycles. There are many factors that affect the magnitude of each of these strains, and consequently the material performance in a pavement, and these are briefly discussed below.

The behaviour of fine-grained soils under repeated loading

The resilient elastic modulus of cohesive, fine-grained soils decreases non-linearly with increasing applied stress, when all other factors are kept constant (Seed et al, 1962). Materials exhibiting high suctions (negative pore water pressures) have been shown to have higher resilient moduli, indicating that stiffness is a function of three stress variables: the confining stress, the axial stress and the matric suction of the materials (Cheung, 1994). Significantly, no changes in resilient response with changing load frequency have been observed.

At low deviator stress levels, permanent deformation has been shown to increase with the logarithm of the number of cycles, the rate of accumulation of permanent strain increasing as the stress increases. This eventually leads to a deviator stress level, denoted the `threshold stress' ($q_{threshold}$), under which the rate of accumulation of deformation increases exponentially. This response has been shown to be related to the material stress history and water content, and thus shear strength. Brown (1996) therefore suggested that the dominant factor in determining permanent deformation is the relationship between applied shear stress (q) and the shear strength of the soil (i.e. stress ratio, q/q_{max}).

Cheung (1994) suggested the concept of a limiting value of q/q_{max} (i.e. $q_{threshold}/q_{max}$) above which plastic deformation increases relatively rapidly. According to this relationship, the accumulation of permanent strain should be approximately linear with the logarithm of the number of load applications for stress ratios that lie below the threshold stress ratio. If q becomes greater than $q_{threshold}$, then the permanent strain accumulates at a markedly increased rate.

The behaviour of granular materials under repeated loading

The principal factor influencing the measured resilient properties of granular materials is the stress level. The modulus (M_r, measured in a triaxial cell) has been shown to increase significantly with increasing confining pressure and slightly with increasing repeated deviator stress, provided that shear failure is not approached.

Permanent deformation in granular materials has been shown to increase proportionally with the logarithm of the number of cycles of applied stress. As with fine-grained materials, a limiting deviator stress below which permanent deformation remains stable has been identified, although research to define the boundary between stable and unstable accumulation of permanent strains in granular materials under cyclic loading, the shakedown concept, continues.

Composite performance of materials in a pavement foundation

Much research has concentrated on defining the factors that affect the individual material behaviour, as described above. However, the interaction between the sub-base/capping materials and the subgrade (composite behaviour) is also of crucial importance for pavement foundation design. The interaction between two layers of material, having different threshold stress ratios and elastic moduli, under transient loading produces a major influence on how each layer and/or type of material can act. If an unbound granular material overlying a softer, weaker subgrade is subject to loading, the deflections will be partially controlled by the upper layer as a result of its load spreading ability (stiffness). This controls the level of stress transmitted to the subgrade. However, the subgrade will influence the amount of load spreading that can take place by the way in which it reacts to the stress that is transmitted. Thus the layer interaction affects the stress distribution, which in turn affects the total elastic and plastic strains that are developed within each material, hence their response to those stresses and vice versa (Fleming and Rogers, 1995). This behaviour ultimately leads to permanent deformation (rutting) within the pavement.

Therefore the measurements of the relevant parameters, both in the field and in the laboratory, should take place under conditions that match as closely as possible those that they will be subjected to *in situ* (i.e. a moving wheel load, see Fleming and Rogers, 1995).

Prediction of environmental changes within a pavement and their modelling

Changes in environmental conditions, in both the short-term and long-term, will influence material performance. While road foundation design is based primarily on construction loading, it is essential to consider long-term behaviour and potential environmental changes, for example where pavement foundations are partially constructed and left for some months before they are completed. In addition, the material characteristics at the construction stage affect the nature of the material's changes, and hence their long-term behaviour, due to the hysteresis effects associated with changes in suction.

Equilibrium water content is reached after the equilibration of (usually dissipation of negative) pore water pressures in the subgrade. This equilibrium value, once attained, remains relatively stable under impermeable pavements. Equilibrium water content has therefore been used for long-term design (Black and Lister, 1979). Factors such as a lowering of the water table (due to the early installation and effectiveness of sub-surface drainage), changes to the stress history of materials (due to the removal of overburden in cuttings or additional stresses due to pavement construction), changes to the material structure (due to the construction operations), material type, temperature, humidity and rainfall may all result in changes to the material's suction (which controls the equilibrium water content), and hence the mechanical performance of the material (Black and Lister, 1979). Although these factors primarily affect the subgrade, the capping and sub-base can also be affected, especially if wet weather occurs during construction.

The effects of changes in suction (hence subgrade water content) on pavement performance are difficult to predict. In the UK, seasonal changes to the level of the water table have been shown to influence the strain response of clay subgrades under load (Sha'at et al, 1992). This is normally of little significance for a full pavement construction as long as the granular layers are adequately drained and 'worst case' subgrade design parameters are used. However, it can be important at the construction stage where applied stresses are significantly higher, and hence lower subgrade stiffness pertains. To enable this effect to be assessed, the changes in water content must be predicted.

The problems of (highly) variable weather and predicting sub-surface water regimes, and the consequences of these on material response, make it difficult to anticipate the long-term changes to the performance of unprotected or partially completed pavement foundations. In addition, measurements of subgrade performance made during construction will be influenced by temporary changes caused by drying and re-wetting, as well as surface remoulding (under tyre action), re-grading and re-compaction.

For accurate laboratory testing, the subgrade condition must be modelled allowing for changes in its compacted state, environmental conditions (hence water content) and applied loading (depending on the construction operations performed), and its location (cutting or embankment). There are four main material states that should be considered:

- Undisturbed: as found in the base of cuttings or 'at grade' at the time of construction.
- Remoulded: re-compacted soil at the in-situ water content, as found in embankments at the time of construction or after reworking.
- Samples in the two conditions above, but at their long-term equilibrium water contents after equilibration.

For laboratory testing, undisturbed samples may be prepared directly from the subgrade, while remoulded samples can be straightforwardly prepared by re-compacting a sample of the subgrade using appropriate compaction methodologies. To create samples that accurately represent the equilibrium condition, however, the prepared sample (either undisturbed or remoulded) should be allowed to change water content under the equivalent conditions of confining stress and suction that would be experienced in the field. Methods of forcing water into pre-compacted (i.e. remoulded) samples have been proposed (Drumm et al, 1997). This revealed, even for high permeability soils, that the soil requires a considerable time to equilibrate. UK subgrades are typically low permeability clay soils and for a commercial test, which would be required to provide data for a performance-based design, such a lengthy procedure would prove impractical. The difficulty of bringing undisturbed samples to equilibrium would be greater. Therefore, methods to predict long-term subgrade water content and then modelling it presents a significant research challenge.

Laboratory and field testing of performance parameters

Laboratory assessment techniques

Sophisticated Repeated Load Triaxial Test (RLTT) procedures with on-sample instrumentation have been developed to assess both the stiffness and permanent deformation behaviour of clay soils. However, due to the nature of the apparatus required, such testing has so far been limited to research laboratories. The RLTT has been suggested as the most appropriate apparatus currently available to assess the pavement performance parameters for design. Whilst this form of testing does not model the true loading experienced under a rolling wheel (i.e. rotation of principal stresses), and is limited to cycling the deviator stress, stiffness is unaffected by this. The permanent deformation behaviour is defined by determining a 'threshold' stress ($q_{threshold}$), below which the development of permanent deformation remains stable (i.e. accumulates at an ever-decreasing rate). Permanent deformation of the subgrade can therefore be controlled by limiting the applied vertical stress, transmitted through the overlying layers, to a level *below* which the accumulation of permanent deformation remains stable.

The performance of granular materials can also be determined using the RLTT, although routine laboratory assessment of materials of the size and nature of those used as capping is currently impractical, due both to the large particle size and the complicated cyclic loading required to simulate traffic loading. This is an area requiring further research. Currently laboratory testing of capping materials is essentially limited to physical index and chemical tests to guard against particle degradation under trafficking and adverse effects of water content changes in the long-term, and to the assessment of compaction properties (MCHW 1998). In addition, assessment of the compaction behaviour of materials containing particles greater than 80mm in size is problematic (Rockliff, 2000). Laboratory based dynamic plate testing to assess stress dependency, similar to that proposed for use in the field, may offer a way forward for the assessment of design parameters for such materials (Fleming et al, 2002).

Field assessment

Considerable research has recently been undertaken to develop dynamic stiffness measuring devices that can quickly measure the stiffness of the subgrade and the pavement layers during construction. These devices measure a composite stiffness under a transient load pulse, which is applied to the ground by dropping a weight onto a bearing plate via a rubber buffer. The deflection of the ground is measured and combined with the applied load, which is either measured or is assumed to be constant (by means of a constant drop height), to calculate the stiffness using conventional Boussinesq static analysis. Such devices include the trailer-mounted Falling Weight Deflectometer (FWD) and a range of portable devices such as the prototype TRL Foundation Tester (TFT), German Dynamic Plate (GDP, sometimes termed the lightweight drop tester), and the recently developed Prima. The portable devices typically apply a stress of 100 to 200kPa via a 300mm diameter plate over a period of approximately 20 milliseconds, and are suggested to be more suitable for testing subgrade and capping than the FWD. The three portable devices measure deflection via a central geophone (or velocity transducer) only, thus assessing only the composite stiffness of the foundation and precluding individual layer stiffness determination by backanalysis. However, by testing each layer as it is constructed their contribution to the overall pavement performance can be indirectly assessed.

It is proposed that the laboratory-derived threshold stress of a soil can be linked to its undrained shear strength (q_{max}, Frost et al, 2002), and that this can then be used for correlation to shear strength measurements on site. The indirect measurement of strength in the field is possible using the portable Dynamic

Cone Penetrometer (DCP). However, the DCP is not suitable for penetrating very strong materials or those containing very large particles, and only measures the properties of the individual material layers as they are being penetrated (i.e. it cannot measure the composite foundation performance).

The mechanisms that lead to the development of rutting in two-layer systems are not well understood, and the ability to model rutting is consequently lacking. Therefore, it is proposed that the only currently practical means of guarding against excessive permanent deformation is to monitor the development of surface rutting in the "as-constructed" road foundation and to place a limit on the degree of rutting allowed. As a consequence, a pre-construction trial section should be constructed and subjected to controlled trafficking to prove the competence of the proposed materials and methods.

Although density is not a performance parameter, its measurement is considered important for assessing the adequacy of the compaction of unbound materials. A material, once laid within a pavement foundation, may possess sufficient stiffness to allow the adequate compaction of the subsequent layers, but may deform excessively during trafficking because of poor strength due to inadequate compacted density, and therefore particle interlock. It is consequently recommended that density should be measured on site and compared to the maximum density that is achievable from either a laboratory test or in the pre-construction field trial. Measurement of dry density is not uncommon in practice, the Nuclear Density Gauge (NDG) being an accepted method.

Specification approach adopted

It is clear from the above discussion that it is not currently possible for the complete requirements of a fully analytical performance specification to be implemented, due primarily to the lack of widely available commercial, and practical, equipment that is suitable for tests of such complexity. The authors, in partnership with the UK Highways Agency, are currently undertaking research to develop such tests. Therefore until suitable tests are available, nicely judged compromises are necessary. In addition, it is acknowledged that the performance specification developed would need to undergo a phased introduction into practice so that experience can be gained of the proposed test methods and data produced. This is considered necessary both to engender confidence in the new approach and to make best use of the considerable empirical experience that has been generated over many years with the traditional method specification.

As a consequence the performance-based specification produced from the initial research, for implementation purposes, currently accommodates two different approaches.

- A CBR performance-based approach, to assess the subgrade, with the traditional two phase design process (short- and long-term) and the new elements of field compliance testing *(standard approach).*
- A fully analytical performance-based approach substituting a new laboratory soil characterisation test to replace the CBR test (*detailed approach*).

The detailed approach is regarded as a longer-term goal, based on the concepts detailed above. Currently, only the standard approach is being evaluated for implementation and is described in detail below.

The performance-based specification (standard approach)

The specification developed features three iterative stages: design, a pre-construction field trial, and construction compliance testing. For the final stage, target performance values are required.

Target values for compliance testing

To ensure that the subgrade properties found in the field are as good as, or better than, those assumed in the design it is proposed that CBR measurements are made in the field and must at least match the long-term design CBR values (the short-term design being the contractor's responsibility). These values will be design/site specific, and necessitate site compliance testing.

To ensure adequate performance of capping during construction, it is proposed that various specification targets are set. A target composite capping stiffness of 50MPa (measured with a 300mm-diameter dynamic plate test) is proposed to facilitate adequate compaction of the sub-base. The dry density after compaction should be at least 95% of the laboratory maximum dry density using the BS compaction test. A limit on the surface rutting of 40 to 50mm (arising from construction vehicles) is proposed to protect the subgrade. The values have been chosen on the assumption that approximately 50% of the capping surface rut is transferred to the surface of the subgrade. These targets and limits are proposed values only, and are being assessed in the implementation work.

Design

The design requirements for the standard approach are similar to those for the existing method specification (DMRB Vol.7 HD25/94, 1994) for the in-service full pavement, though resulting in lower capping requirements than at present. This long-term design requirement utilises the CBR (based on prediction of the equilibrium water content in accordance with current guidance). However, for the short-term design of the foundation (it being the contractor's responsibility to ensure that the foundation can be built and provide an adequate platform for the construction of the upper structural layers) the CBR, or an alternative parameter such as stiffness, can be utilised. The design thicknesses are established for these two conditions using charts which will be included in the new specification. These are based on static linear elastic theory and are for guidance only for the short-term case, but are compulsory for the long-term case; the greater of the two thicknesses should be chosen.

Field trial

To demonstrate that the selected materials and design are adequate, a site trial is to be performed prior to construction using the proposed materials and methods on a *representative* section of subgrade. A complete programme of in-situ testing is to be performed on the subgrade and capping to determine the performance of the trial relative to the design data and target values (i.e. measurement of CBR, stiffness, strength and density). The trial section will then be trafficked to determine its resistance to permanent deformation by monitoring the rutting of the capping surface. In the event of large surface ruts occurring, the proportion of rutting transmitted to the subgrade surface can be determined by excavation, and any adjustment to the specified rut limits agreed on a site-specific basis. At this stage the contractor can consider different thicknesses and combinations of materials to optimise his design.

Once the standard testing is complete, consideration should be given to artificial saturation of a limited area of the trial section, particularly if the materials are considered to be seriously water susceptible, followed by further assessment to examine the possible effects of poor weather during construction. If the trial proves unsuccessful, the design thicknesses and/or choice of materials must be re-evaluated and a further trial should be carried out. Finally, after successful confirmation of the design and the specification values from the trial, construction can begin.

Construction

The subgrade is to be tested *in situ* immediately prior to capping placement to check that it meets the design values for the long term. However, if the parameters measured on the subgrade lie below the long-term design values (or any other pre-determined values, based on laboratory test data, which suggest that the equilibrium values will subsequently fall below the design values), then the long-term design may need to be amended as construction takes place. 'Soft spots', if these are the problem, will need to be isolated and treated accordingly. If the targets for the subgrade fall below the short-term requirements for site construction (this is the contractor's responsibility, and thus the contractor must balance the construction costs versus risks for these situations), then either additional excavation and addition/thickening of capping or subgrade stabilisation may be needed. If either of these solutions is adopted, it may be possible to take account of this in the long-term design and reduce the thickness of the sub-base, or even pavement layers, accordingly. However, this course of action needs careful consideration and is not presently covered in the specification.

Once the subgrade is shown to be acceptable, the capping can be constructed. The amount of surface rutting under construction trafficking should be monitored as construction works proceed and compared to the limiting values. Capping density should be checked to guard against long-term deformation, and the top of capping composite stiffness measured immediately prior to sub-base construction to ensure that adequate compaction of the sub-base can be achieved.

Implementation trials

Evaluation and refinement of the draft performance-based specification is now taking place. The development of the draft specification was based on many measurements at several 'live' construction sites and purpose-built full-scale 'controlled' field trials. Consequently, the implementation work has focussed on how the proposed specification fits in with the many different forms of contract, and its impact on the construction operations, standard testing regimes and general project management procedures. In order to evaluate the difficulty of implementing the specification, several live sites using different forms of contract have been tested and this work will be published in due course.

Implications of a performance approach

In the near future, it is considered that the proposed change to a performance-based specification approach will not significantly affect the use of traditional materials, but will open up new possibilities for other materials. The current specification clauses for capping materials will remain largely unchanged. However, the new specification will allow contractors to use a wider range of materials, if their performance can be demonstrated to be acceptable. The most significant change will be that materials provided will have to be shown to be able to perform *in situ*, i.e. that the materials can be trafficked (as per the site requirements) without excessive rutting and that a target stiffness can be achieved when compacted onto a typical subgrade. Therefore a greater appreciation of the performance of supplied aggregates will be needed from their suppliers and contractors in general.

In the medium-term it is anticipated that the material suppliers will be required to provide performance data relating not only to the durability of their materials but also the performance parameters of stiffness and permanent deformation. Similarly, constructors will be required to provide assurance of any proposed material's suitability and performance once placed.

In the longer-term the move towards a fully analytical approach to pavement foundation design will require a much greater understanding of both the performance of the materials supplied and the use of appropriate performance test methods. In addition, the performance of stabilised materials has to date been investigated to a lesser extent than unbound foundation materials, and this is clearly an area requiring further research.

Conclusions

The current CBR-based recipe specification for pavement foundations is restrictive and impedes the use of recycled and marginal materials. It also restricts the full benefits of analytical pavement design procedures from being implemented.

Recent advances in laboratory and in-situ testing have enabled a move from a recipe specification to a performance-based specification for road foundation materials. The parameters that need to be assessed, both for design and proof of performance on site, are resilient elastic modulus (or stiffness), shear strength and resistance to permanent deformation.

To ensure adequate field performance, the strength and stiffness of the subgrade should be tested prior to placement of capping and/or sub-base. Similarly, the finished capping and sub-base surfaces should be tested for composite resilient elastic modulus to confirm that the design requirements have been met. It is not currently possible to measure directly a parameter that indicates the composite resistance to surface deformation and consequently the accumulation of surface rutting should be monitored and kept below a limiting value, dependent on the thickness of the combined capping/sub-base.

A range of testing devices exists for the assessment of pavement foundation layers *in situ*. It is recommended that a pre-construction trial section is constructed to establish correlations between laboratory and field measurements, as well as to prove the efficacy of the proposed construction materials and methods used. It is proposed that trafficking of the trial section be carried out so that the likely performance can be accurately determined. The pre-construction trial should give both the client and the contractor confidence that the works will be constructed to an adequate standard, to ensure future performance.

A comprehensive set of trials on live construction sites and specifically-constructed trial road foundations has been carried out in support of the specification philosophy developed. Detailed recommendations for the assessment devices and procedures to be used and the target values that need to be achieved have also been suggested and are published elsewhere. This work has proved that the philosophy contained herein provides a sound and fair basis for the specification of pavement foundations, whether constructed from traditional materials under traditional loadings regimes or novel materials under revised loading.

Further development of laboratory testing techniques is required to produce simple and routine measures of the required performance parameters for both the subgrade and overlying foundation materials. The main issues are associated with the testing of large particle size granular materials, stabilised materials and the prediction and modelling of long-term environmental conditions within the pavement. Research is underway to address these issues.

When the precise values of a material's stiffness and resilient deformation can be measured routinely, a framework has been created that can incorporate the measures in the specification using analytical design procedures. This will allow freedom to use materials which are currently excluded and potentially may allow changes to the design of the structural pavement layers as a full and accurate consideration of the foundation properties can be made. This is of particular significance for stabilised subgrades that frequently have better properties than the imported sub-base materials that are specified to overlie them.

The performance specification approach adopted can clearly be transferred to any type of pavement or infrastructure construction that relies on measured or inferred values of CBR (as an assessment of strength and stiffness). Therefore the approach should be suitable for the assessment of structures such as heavy-duty paving and floors, trench reinstatement and rail trackbed design, as well as minor and major road construction or repair.

Acknowledgements

The authors are grateful to Scott Wilson (PE) Ltd. and Nottingham University for their collaboration on this work, and to the UK Highways Agency for funding the research. The opinions expressed are those of the authors and not necessarily those of the UK Highways Agency.

References

Black, W.P.M. and Lister, N.W. *The Strength of Clay Fill Subgrades: Its Prediction in Relation to Road Performance*, Transport and Road Research Laboratory Report LR889, Crowthorne, Berkshire, UK. 1979.

Brown, S.F. *Soil Mechanics in Pavement Engineering*. 36th Rankine Lecture of the British Geotechnical Society. *Geotechnique*, Vol. 46, No. 3, pp. 383-426. 1996.

Cheung, L.W. *Laboratory Assessment of Pavement Foundation Materials*. Ph.D. Thesis, University of Nottingham, Nottingham, UK. 1994.

DMRB : HD 25/94, *Foundations*. In: *Design Manual for Roads and Bridges*, Volume 7. HMSO, London, UK. 1994.

Drumm, E.C., Reeves, J.S., Madgett, M.R. and Trolinger, W.D. *Subgrade Resilient Modulus Correction for Saturation Effects*. Journal of Geotechnical and Geoenvironmental Engineering, ASCE, Vol. 123, July, pp. 663-670. 1997.

Fleming, P.R. and Rogers, C.D.F. *Assessment of Pavement Foundations During Construction*. Transportation, Proceedings of the Institution of Civil Engineers, Vol. 111, No. 2, pp. 105-115. 1995.

Fleming, P.R., Lambert, J.P., Frost M.W. and Rogers C.D.F., *In-Situ Assessment of Stiffness Modulus for Highway Foundations during Construction*, 9th International Conference on Asphalt Pavements, Copenhagen, Denmark, August, 2002.

Frost, M.W., Fleming, P.R. and Rogers, C.D.F, *Threshold Stress and Asymptotic Stiffness of UK Clays in the Repeated Load Triaxial Test*, Proceedings of the 6th International Conference on the Bearing Capacity of Roads and Airfields, Lisbon, Portugal, Vol. 2, pp 1099-1108. 2002.

MCHW Vol.1, *Specification for Highway Works*. Manual of Contract Documents for Highway Works, Vol. 1, HMSO, London, UK. 1998.

Powell, W.D., Potter, J.F., Mayhew, H.C. and Nunn, M.E. *The Structural Design of Bituminous Roads (LR1132)*. Transport Research Laboratory, Crowthorne, Berkshire, UK. 1984.

Rockliff, D. *European Standards for Unbound Mixtures - A Status Report*. In Proceedings of the 5th International Symposium on Unbound Aggregates in Roads, Nottingham, pp 49-54. 2000.

Seed, H.B., Chan, C.K. and Lee, C.E. *Resilience Characteristics of Subgrade Soils and Their Relation to Fatigue Failures in Asphalt Pavements*. Proceedings of the International Conference on the Structural Design of Asphalt Pavements. University of Michigan, Ann Arbor, USA, pp. 611-636. 1962.

Sha'at, A.A., Kamal, M.A. and Matter, N.S. *Relationship between Climatic Conditions and the Structural Parameters of Flexible Pavements*. Proceedings of the 7th International Conference on Asphalt Pavements. International Society for Asphalt Pavements, Nottingham, UK, Vol. 4, pp. 326-340. 1992.

Recycling in transportation

S.J. Ellis
TRL Limited

Introduction

Recycling as an activity is one of the fundamental elements towards achieving sustainability. Sustainability has been identified as 'the capacity for continuance into the long term future'. Anything that can continue being done on an indefinite basis is sustainable, the practices of the world's populace are currently unsustainable. If a point is not reached where activities are sustainable, then eventually all available resources will be consumed and a level of pollution will be generated that will mean the earth is no longer capable of sustaining human life. The challenge is to achieve social progress and economic growth without resource depletion and adverse environmental impact. For the construction industry, for example, it is clear that the continued quarrying of primary materials (Figure 1) is unsustainable hence recycling becomes essential if growth is to be maintained.

The Government objective for our transport system, giving due consideration to sustainability, is that it should provide the choice, or freedom to travel, but minimises damage to the environment. In designing for construction projects the way in which materials are specified is changing to allow for innovation and alternatives to the use of primary aggregates newly extracted from a quarry. For instance, specifications can be based upon performance, rather than on the use of standard materials made to strict recipes. Design procedures can be introduced to permit many options to be considered, rather than demanding combinations of strengths and layer thicknesses that limit choice to new materials. To be truly sustainable the use of all resources (aggregates, binders, fuel) should be considered alongside engineering requirements in the selection of construction techniques.

This paper provides brief descriptions to define recycling, details potential materials that could be utilised in construction through recycling, and gives examples of their application. Finally, after describing what could be achieved barriers are identified that need to be overcome for recycling to contribute to the sustainable future of the planet.

Figure 1, Quarrying of primary aggregates

Aspects of recycling

Different aspects to recycling in transportation are outlined below, including reference to the way in which these methods can contribute to sustainability.

Recycling and stabilisation

Recycling should now be a major construction activity. In situ, hot, cold, and ex situ techniques are now all feasible and many large and specialist contracting organisations can offer these services.

In situ recycling occurs when the existing materials are treated on site to be used in a new structure, the most obvious being in new pavement structures. There are possibilities for all layers of the bound pavement structure to be recycled in this way.

Cold in situ recycling for the structural maintenance of highway pavements has been developed and used in the UK, TRL Report 386 (Milton and Earland, 1997) provides a specification and design guide that considers deep cold in situ recycling. The advantage when using cold mix techniques is that less energy consumption is required than for hot mix methods.

A further in situ recycling technique is that of ***soil stabilisation*** by means of application of lime and/or cement or some industrial by-products to treat the sub-grade to pavement structures. By carrying out this technique the need for substantial pavement foundation layers is removed and therefore the volume of imported materials is less.

Ex situ recycling differs from in situ recycling in that the materials are removed from the site to be used as aggregate in plant mixed materials. Although this still makes good use of materials it can increase vehicle movements above those required for in situ recycling.

Alternative materials

Aside from recycling existing conventional construction materials, there is now a wide range of alternative materials also available for use. These materials may have properties different from primary aggregates because of their mode of formation, and this affects their engineering behaviour. In many cases, these properties can be turned to advantage to create new materials that have no natural analogues, such as the use of pulverised fuel ash or ground granulated blast furnace slag as hydraulic binders. In other cases, alternative materials may be used in place of primary aggregates to preserve scarce natural resources. Alternative materials that can now be readily used in highway construction include: construction and demolition waste, asphalt arisings, pulverised fuel ash (pfa), china clay sand, slate waste, steel and granulated blast furnace slag (ggbfs), colliery spoil, incinerator bottom ash (iba), crushed glass, waste tyres and waste plastic. Table 1 provides a comprehensive list of materials available in the UK, their location, quantities available, and potential uses.

Examples of recycling for transportation geotechnics

There are some recycling applications that are commonly recognised, two particular products in this category are the use of pfa and asphalt arisings in construction. Pfa for general fill or in structural concrete is virtually considered to be the 'norm'. The availability of asphalt arising is considered to be scarce as the product is already re-used in many forms, ranging from patching farm tracks to equivalent Type 1 granular sub-base material for trunk roads. In addition, in continental Europe asphalt arisings are as aggregate in structural concrete. The ready use of these by-products is encouraging though as discussed later for the future of recycling in the UK there is scope to improve the application situation for some recycled products and achieve higher value returns on their use.

Aside from the established recycling previously mentioned, ten years ago there were few examples to call upon to demonstrate the application of the many other recycling techniques in construction. This situation has changed due to the fiscal drivers imposed by Government, the landfill tax and aggregate levy, promoting increased innovation by suppliers and contractors to meet the demands for a sustainable future. Figures 2 and 3 and the following Table 2 provide a sample of the applications of recycling for geotechnical engineering.

Table 1, Alternative materials, their availability and current potential in the UK as of 2001 (after, Reid and Chandler, 2001, note nd=not determined)

Material	Arisings (Mt/a)	Stockpile (MT/a)	Location	Uses
Baghouse dust	nd	nd	Asphalt plants	Filler in asphalt
Basic oxygen furnace steel slag (BOS)	1.0 (98% used)	No reliable quantitative estimates	N and E England, S Wales	Surface dressing Bound in roads (cement, bitumen or hydraulic) Unbound in roads Concrete aggregate
Cement kiln dust	0.13 (4 works)	nd	Cement works	Filler in asphalt
China clay by-products	22.6	600	Cornwall, Devon	Bound in roads (cement, bitumen or hydraulic) Unbound in roads Concrete aggregate (fine) Artificial islands Embankment dams Most SHW specifications Coarse Concrete aggregate
Coal and other mining waste (colliery spoil)	7.5 (coal only)	10-20 useable 3600 hard-to-use	Coalfields, mineral workings	General Fill in roads (unburnt) Selected granular Fill and Capping and sub-base, either unstabilised or stabilised (burnt)
Construction and demolition wastes	93.91 Range 79.8 - 108	0	Principally urban areas	Bound in roads (cement, bitumen or hydraulic) Unbound in roads Concrete aggregate Artificial islands General Fill Grout backfill to mine workings embankment dams
Crushed brick	5	not recoverable	Demolition sites	Hardcore Granular sub-base Recovered bricks
Foundry sands	0.9	nil	West Midlands & Yorkshire	Fine aggregate in Concrete blocks Bricks Road sub-base Asphalt Foamed Concrete

Material	Arisings (Mt/a)	Stockpile (MT/a)	Location	Uses
Glass cullet	2.2	nd	Municipal waste	Aggregate in asphalt also as Fill Unbound in roads Sand in concrete Drainage filter media
Municipal solid waste incinerator bottom ash	0.64 (1.35mt by 2003, and 2.5 mt by 2010)	Negligible	Generally urban areas	IBA has been used successfully in Europe as: • embankment Fill; • aggregate for bound layers in roads ; • aggregate for building blocks; • daily cover material for landfills.
Power station fly ash (PFA)	4.9	55	Coal-fired power stations	Cement extender in Concrete General Fill, lightweight Fill Capping Sub-base Road base Concrete additive Manufacture of lightweight aggregate
Quarry fines	nd	nd		
Quarry waste (non-fines)	nd	nd		Bricks Lightweight aggregate Autoclaved Concrete blocks
Scrap tyres	0.4	No estimates available. Hard-to-use long term stockpiles 116.9 kt		Asphalt French Drains Cement kiln fuel Sports and playground surfaces Compressible fill to bridge abutments
Slate waste	6.33	465.5	Wales (97%), Scotland, Pennines	Low grade aggregate Concrete Horticultural soil improver Type 1 bulk Fill Selected granular Fill and Capping Manufacture of lightweight aggregate

Material	Arisings (Mt/a)	Stockpile (MT/a)	Location	Uses
pent oil shale	0	100	West Lothian	General Fill in roads, relatively lightweight Selected granular Fill Capping and sub-base, either unstabilised or stabilised
Spent railway track ballast	1.3 0.005	Negligible. No estimates available		Railway track ballast Concrete
Trench arisings	nd	nd		
Waste glass	5% of vehicle & demolition waste			
Waste plastic	2.8			Pipes? Bearing and sheet piles? Geotextiles

Table 2, Summary examples of recycling for geotechnical engineering

Recycled materials	Geotechnical application	Location	Date	Reference
Spent Oil Shale (Figure 2)		Central Scotland	1960s onwards	Winter 1999
Secondary aggregates and binders	Pavement foundations	TRL trials	1997 onwards	Atkinson and Chaddock 1999
Colliery Spoil (burnt and/or unburnt)	Embankment Fill Capping Sub-base	UK		Winter 2001
Railway track materials	Ballast (Figure 2)	UK	1998 onwards	Reid and Chandler 2001
Highway maintenance arisings	Capping, Embankment Fill	Bemersley Tip access road	1999	Reid and Chandler 2001
Sub-grade Oil shale aggregates	Lime / PFA stabilisation PFA/cement bound sub-base	The Echline experimental road	1999-2000	Reid and Chandler 2001
Contaminated materials Sub-grade	Environmental bunds Lime stabilisation	M60	2000	Reid and Chandler 2001
Granular material (IBA) treated with Fly Ash (GFA) Demolition waste	Sub-base Capping	Burntwood Bypass	2001	Reid and Chandler 2001
Asphalt, concrete & brick arisings	Sub-base Capping General Fill	Nottingham Express Transit	2001-2002	Reid and Chandler 2001

Figure 2, Recycling railway ballast

Figure 3, The Five Sisters spent oil shale West Lothian. (Recent estimates indicate that around 100,000kt of spent oil shale is stored in similar tips in the eastern central belt of Scotland)

The future for recycling in the UK

Winter (2002), has described a conceptual framework for the recycling of aggregates and other wastes. The framework was developed through a study of recycling activity in Scotland. The framework is described in terms of both the environmental and economic utility of the recycled application and the use relative to the original application. The framework for use is set-out in terms of low, intermediate and high utility and considers relative use in terms of down-cycling, level-cycling and up-cycling; defined as:

Down-cycling: Recycling in which the secondary application has a lower utility than the primary (e.g., asphalt arisings recycled as general fill).

Level-cycling: Recycling in which the secondary application has the same or similar utility as the primary (e.g., asphalt arisings recycled into bituminous pavement layers).

Up-cycling: Recycling in which the secondary application has a higher utility than the primary (e.g., pavement foundation layer recycled in bituminous pavement layers).

The results of the pilot study in Scotland are demonstrated in Figure 6, which presents the most likely picture of what is happening in the recycling market throughout the UK at the present. The interpretation indicates that the bulk of current applications for recycled aggregates are carried out at low utility and by down-cycling, in general lowering the value of the aggregate with limited recognition of the potential to recycle at a higher value. There appears to be considerable scope and there are applications of recycled aggregates for level-cycling to occur, but there are currently only limited examples of up-cycling.

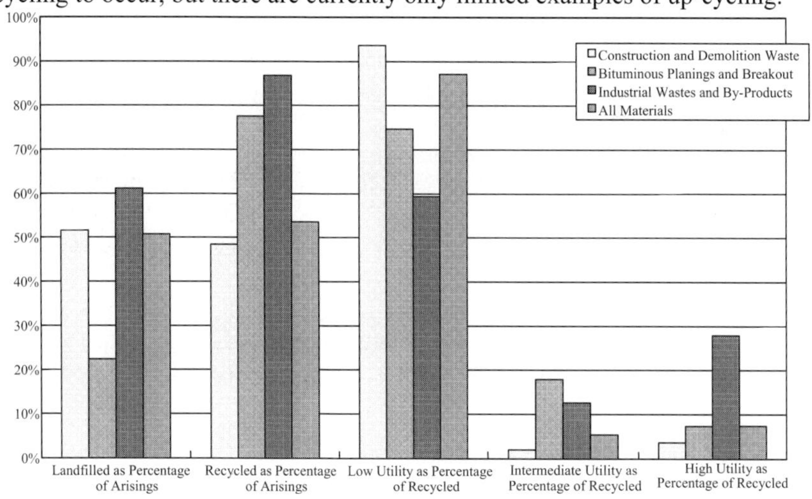

Figure 6, Summary pilot survey data

Table 3, Summary of issues relating to recycling in transport infrastructure, (after, Reid and Chandler, 2001)

Issue	Description	Available Guidance
Specifications	Some materials and methods are excluded from existing Specifications	A number of Specifications for alternative materials and methods are available
Test methods	Existing test methods developed for natural materials are not7 suitable for some alternative materials	A number of tests have been assessed as suitable for alternative materials
Reliability and Quality Control	Alternative materials perceived as highly variable and of low quality	Utilise or adapt existing quality control systems to produce a consistent, fit-for-purpose material
Environmental Concerns	Potential long term leaching of contaminants into controlled waters; dust and noise during construction	Assess behaviour using leaching tests and existing models where necessary; CDM/COSHH legislation
Waste Regulations	Unclear whether materials are waste or covered by exemptions, potential long time scale required by waste permitting processes	Use available guidance on waste permitting system
Conditions of Contract	Some forms of contract may create an environment where there is no incentive for innovation	Use appropriate forms of contract and adopt partnering
Planning	Difficulties getting planning permission for recycling centres in or near urban areas	Guidance for planners and applicants has been produced by DETR
Supply and Demand	Difficulty in matching supply and demand for some alternative materials	Plan in advance and stockpile material if necessary; use existing databases to source materials
Economics	Alternative materials and methods may be more expensive than conventional ones	Ensure comparing like with like; use whole life costing to ensure best practical environmental option selected
Lack of Awareness	Many individuals and organisations unaware of the possibilities, or only aware of potential problems	Disseminate existing information from CIRIA, EA, TRL, BRE, AAS and others

This paper has so far identified the positive developments that have and are taking place, in order to implement sustainability. Nevertheless, despite all the progress that has been made, introducing sustainability into all sectors of the construction process, is not a straightforward process. A DTI Partners in Innovation project 'Recycling in Transport Infrastructure', www.viridis.co.uk, went some way (see Table 3,) towards identifying the issues and barriers associated with recycling.

Since these issues were identified there have been many attempts by Government and industry to address them. In particular the Government does sponsor a substantial volume of research and dissemination projects to promote recycling. Through one such project, the DTI have sponsored a dissemination programme for sustainable construction in practice (Ellis 2003). The documents available through the web-site cover many of the issues listed and in particular, Reid (2003) has collated the available sources for specification and quality control when using alternative materials in construction. Much guidance has developed in recent years, and increased awareness of both the materials available and the specifications available for their recycled use should lead to an increased uptake of recycling in the construction industry and a more sustainable future in the UK.

Conclusion

Recycling is a fundamental part of the future sustainability of our planet. The potential for recycling within the construction industry is great and can contribute substantially to the longevity of the planet. Much progress has been made in the UK and it is important that the developments achieved in research are successfully implemented through industry. The recycling topic is vast and this paper provides a brief review of what could be achieved, the references provide more detailed information on recycling and the engineering solutions provided by alternative materials. By increasing the volume of recycling within construction, simple 'sustainability' advances can be made, for example, through direct savings in fuel consumption (vehicle movements and power supplies), or savings in material consumption (pre-fabrication, material sorting and control). Continued dissemination to industry will ensure that more complex advances can be made and organisations will be able to turn recycling into an economic advantage.

Acknowledgements

The author is grateful to Dr J M Reid (Viridis), JWE Chandler (TRL Limited) and Dr MG Winter (TRL Limited) for their contributions to this paper.

Any views expressed in this paper are those of the author. Whilst every effort has been made to ensure that the matter presented in this paper is relevant, accurate and up-to-date at the time of publication, TRL Limited cannot accept any liability for any error or omission.

References

Atkinson VM and BCJ Chaddock. *Enabling the use of secondary aggregates and binders in pavement foundations*. TRL Report 408. 1999.

Chaddock BCJ and VM Atkinson. *Stabilised sub-bases in road foundations; structural assessment and benefits.* TRL Report 2481997

Ellis SJ. *Construction materials*. Sustainable Construction in Practice Roadshow published paper & presentation. Also published www.trl.co.uk/scip. 2003.

Milton and and MG Earland. *Design guide and specification for structural maintenance of highways by cold in-situ recycling.* TRL Report 386. TRL Limited, Crowthorne. 1999.

Reid, JM and JWE Chandler. Recycling in transport infrastructure. TRL Limited, Crowthorne. 2001

Reid, JM. Specifications and quality control for alternative materials in construction. http://www.trl.co.uk/scip/delegates/delegates.asp. 2003.

Winter, MG. The use of spent oil shale in earthwork construction. Geoenvironmental Engineering - Ground contamination: pollutant management and remediation, 565-574. London: Thomas Telford. 1999.

Winter, MG. Colliery spoil use in road earthwork construction. Proceedings, Minerals 2001, 1-13. Minerals Engineering Society: Edinburgh, 2001.

Winter, MG. A conceptual framework for the recycling of aggregates and other wastes. Proceedings of the Institution of Civil Engineers (Municipal Engineer), 151(3), 177-187. 2002.

Hydraulically bound soil for road foundations

S. Biczysko
Atkins Highways and Transportation

Introduction

The foundation for a road or paved area has two principal objectives:

- to resist the relatively large stresses and strains associated with construction, and
- to distribute in-service stresses and strains generated by the traffic using the road throughout its functional life.

Foundation stresses during construction are relatively large but are limited in number. The in-service stresses are small, due to the presence of the road layers above the foundation, but accumulate to a very large number of load pulses over the life of the road generated from the traffic spectrum using the carriageway. A key facet of the foundation is good maintained functional characteristics to ensure that the performance of the whole road structure is not compromised. The purpose of the road structure is to transmit the dynamic stresses from the tyre contact area at the road surface through the road pavement and into the soil in such a way as to ensure that the soil is not subject to overstrain. These are a number of differing ways that this can be achieved by selection of appropriate materials and thicknesses.

Historically, roads have evolved through the ages from what was originally unbound granular material. The unbound aggregate layer is today frequently used as the foundation for a modern road and forms the transition between the soil beneath and the bound layers of the road above. Whilst unbound material provides a traditional foundation medium, its elastic modulus, i.e. efficiency of load spreading, is only of the order of three times that of the underlying soil and if water or groundwater finds its way into the unbound layer its effectiveness can be compromised through inability of good drainage and displacement through frost action.

Many of the United Kingdom soils are fine grained and cohesive materials. Design input indices, in the form of equilibrium CBR, have developed and been refined over many years to generate the thickness of foundation layers. The essence of this index ratio is the propensity of the soil to suction effects and volume change as characterised by Plasticity Index. It has become established practice to use unbound aggregate as the foundation above the soil, either in its entirety (sub-base) or as a two layer concept (sub-base and capping). In the two layer foundation structure the lower layer can be of a less demanding 'quality' than the upper level. Although granular materials predominate in both cases there is an array of differing materials which can be used to fulfil the engineering functions of the foundation.

In essence, the performance of the road structure is an increasing value elastic modulus progressing from the subgrade through the base and binder course of the structure. Although unbound aggregate has become the general material to fulfil the foundation objectives there are alternative materials and processes which can equally, or even more efficiently, provide engineering purpose. The focus for this has become greater over the last few years with the drive towards more sustainable construction, conservation of resources and the influence of climate change.

The concept of removal of soil to replace it with a granular material to form a road foundation is a practice which, in a sustainable environment, must be considered carefully. The characteristics of the soil itself can be markedly altered and enhanced to deliver engineering characteristics which will fulfil the requirements of foundation performance.

For some time the use of lime as a modifier of soil has been an adopted practice and this is extensively described by Rogers et al (1996) and others in the technical press. Lime treated soil is an option available in standard UK design and construction. However the use of lime stabilised soil is generally at the lower levels of the foundation as an alternative capping material. The whole of the foundation could be formed of modified soil albeit that other hydraulic binders would be required as complimentary partners to the lime component. This could provide the opportunity for the hot-mix asphalt base material to be paved directly onto the modified soil with a granular sub-base not required at all and the whole foundation formed from hydraulically bound soils.

Lime stabilisation

A substantial number of highway schemes have been constructed using lime stabilisation of the existing soil as a substitute lower level for the foundation. The vast majority of these projects have resulted in durable and stable road pavement structures. Soil chemistry is a key factor in the engineering design of lime stabilised solutions. Several soils contain sulphate, or sulphur which could oxidise to sulphate after exposure and disturbance. The presence of sulphate can lead to expansion and swelling linked to ettringite formation and in some cases considerable volume change can occur. Since many of the soils in the UK can contain sulphates careful attention needs to be given to this facet at the time of site investigation if robust lime stabilisation solutions are to be developed. For the principles of soil stabilisation to be adopted as the total foundation structure, and not just the 'capping' level, for a highway there would have to be full surety that sulphates were not present in any quantity which posed measurable risk. Alternatively, and more realistically, the lime component would need to be blended with a second hydraulic binder to guard against the risk of prospective expansion during the service life of the road foundation.

Hydraulic bound foundation

During the mid 1990s interest become directed towards the practicability of foundations constructed with two hydraulic binders to transform clay soils into a medium which would provide a stable and durable foundation platform onto which hot-mix asphalt materials could be laid and compacted. A temporary extension of the A421 road at Tingewick provided an extended performance test for the concept and has been described by Higgins and Kennedy (1999). The temporary diversion section of road, which remained in service for over one year, provided a practical testbed to evaluate combinations of lime with ground granulated blast furnace slag (ggbs) and also lime with Portland cement. These full scale trials were constructed in the works as a series of experimental bays with differing combinations of soil and binders. The parent soil was Glacial Till (Boulder clay) with Plasticity Index of 28 (I_p). The stiffness of the hydraulic bound foundation as measured with a Falling Weight Deflectometer (FWD) is illustrated in Table 1 and shows a significant long term increase in stiffness modulus over elapsed time.

Table 1, Time related stiffness (MPa) as measured by FWD (after Higgins and Kennedy, 1999).

Soil plus 1.5% lime plus	8.5% ggbs	6.5% ggbs	8.5% OPC
Pre stabilisation	50	50-60	60
7/8 days after stabilisation	470	420 – 550	590
15 months after stabilisation	700	580 – 850	690
15 Months after stabilisation (with 130mm asphalt above)	1040	950 – 1040	950

The early life strength gain of the stabilised soil during construction was also evaluated using a portable lightweight drop tester. The operational characteristics of this light dynamic measurement device differ from FWD but enables immediate use and application of data on site and provides a direct measure of as-installed dynamic modulus (Evd). This device therefore has the capacity to measure early life strength gain and to confirm that the maturity process is proceeding as required. These effects are illustrated in Figure 1 for the first few hours of the 'maturing' process and can be compared with 'typical'

Evd values in the range 50-70MPa for a crushed rock granular sub-base of equal thickness of 350mm.

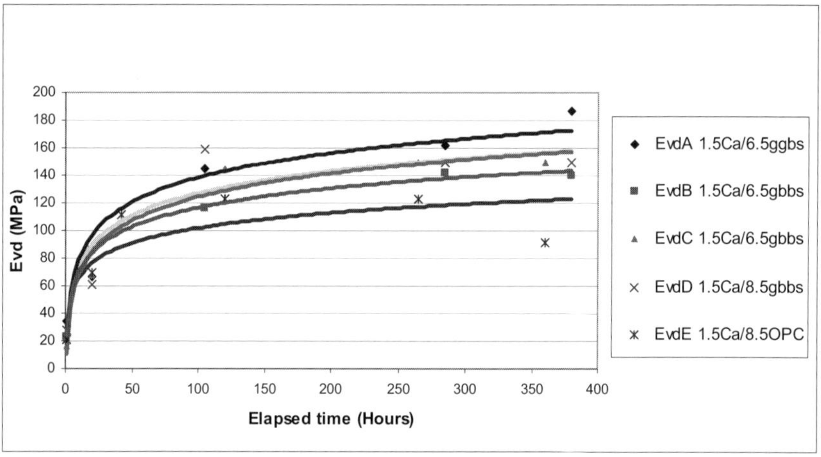

Figure 1, Early life stiffness gain with time, (Ca, quicklime)

The Moisture Condition Value (MCV) apparatus was used on this scheme not only to characterise the acceptability of the Glacial Till soil prior to treatment but also to produce consistent samples of stabilised material for subsequent laboratory test evaluation. The MCV apparatus was found to be a valuable technique in sample preparation since a uniform degree of compaction is provided and the specimen produced in the MCV cylinder can be sub-divided to yield disk shaped specimens suitable for such tests as indirect tensile stiffness modulus in the laboratory.

Hydraulic bound mixture characterisation

All sites will display differing geotechnical conditions and there is an array of test methodologies which can be applied to examine the benefit of stabilising and binding the soil with hydraulic binders. There is also a key factor associated with design where the benefit of hydraulic bound foundation material may be compared with traditional granular material. Chaddock and Atkinson (1997) indicated that significantly greater stiffness modulus can be generated by hydraulic binders with soil as compared with granular material.

Advantage can, therefore, be taken of one of the following options;

- reducing the thickness of foundation, or
- maintaining the same thickness of foundation and reducing the thickness of the overlaying asphalt pavement, or
- retaining the thickness of foundation and overlaying asphalt pavement and taking the benefit in extended functional life of the structure.

Since there can be some degree of variability in natural soil and also in the mechanics of construction the third option above, perhaps presents the optimum selection until greater confidence is gained through serviceability evaluation.

In order to achieve the objective of continuity of measurement the same test devices are best used at each of the key stages:

- Ground investigation (field)
- Mixture characterisation (laboratory)
- Validation (site construction)

As practical and robust tools to determine elastic stiffness and strength the portable Light Dynamic Plate (LDP) and the Dynamic Cone Penetrometer (DCP) were selected. These test techniques can be used at ground investigation stage in exploratory pits, in the laboratory at mixture characterisation stage, and also on site to confirm that design objectives have been achieved in construction.

At ground exploration phase the dynamic plate and cone penetrometer were added to the array of other tests conducted during the investigation fieldwork. These devices yield valuable benchmark data for the natural soil profile by carrying out the tests at sequential levels during the formation of exploratory pit excavations during the site investigation.

(a) (b)

Plate 1, (a) Light Dynamic Plate, and (b) Dynamic Cone Penetrometer tests being carried out in an exploratory pit during a ground investigation

In the laboratory a series of interrelated geotechnical variables can be examined by mixing the soil with various combinations of hydraulic binders and compacting these mixtures into large steel moulds of approximate dimensions 500mm x 400mm x 200mm. The hydraulically bound soil `blocks` are of sufficient size to enable the same dynamic plate and cone penetrometer tests, as used in the ground investigation, to be undertaken in the laboratory to provide values of surface stiffness modulus and layer strength and consistency. The effects of various combinations of soil and binder can also be studied in respect of time related changes in characteristics. Other, more routine, tests are required also and specimens for compressive strength, indirect tensile strength and volumetric swell can be manufactured using MCV apparatus. Density correlation can be established between the refusal density in the MCV specimen and the large block of compacted stabilised soil. All of these data enables evaluation and analysis to derive the most effective foundation solution. The laboratory and ground investigation test information can be supplied with the contract documents to prospective tenders for the works. This avoids the need for all parties who are tendering for the works to glean baseline information in terms of the stabilised soil characteristics during the relatively short tender preparation period and encourages development of bespoke elaboration during the 'window of opportunity' period prior to submission of tender.

During the construction phase the light dynamic plate can be used to validate the works and ensure that the design objectives have been delivered by providing traceability through from the ground investigation and mixture characterisation to the as-installed characteristics in the road foundation.

Specification

The characteristics of soils will differ from site to site but the development of a generic specification for Hydraulic Bound Foundation Material (HBFM) enables a 'standard' method of description. This form of specification must be based upon performance 'outcomes` to enable innovation by the installer of the HBFM but also needs to safeguard the client in terms of durability and functional life. A pathfinder specification clause to deliver these objectives is presented in Appendix I. Since soil characteristics vary from one scheme to another it is necessary to be job specific in terms of some parameters.

The specification addresses:

- Durability through a requirement for a minimum total hydraulic binder content of not less than 8% by mass (ref para 1.5 in Appendix 1).

- Soil acceptability for compaction through a range of MCV. This should be established by relationship testing in the laboratory at the mixture characterisation stage of the HBFM evaluation process. Generally on MCV between 8 and 12 (or 13) should provide an appropriate range of acceptability but this will be dependant upon the nature of the soil and the combination and quantity of hydraulic binders. [ref para 4.3 in Appendix 1] The density in the HBFM layer in the works is required to be not less than 90% of the MCV 'refusal density'. An elapsed period of six hours is described for the field density core cutter specimen to be taken from the treated soil to determine this as the characteristics of the hydraulically bound mixture will be changing with time. Once the relative density correlations have been established with confidence through calibration in the works then density variation can be controlled and monitored by indirect, nuclear, methods.

- Layer thickness and strength through either the use of hand dug exploratory pits to enable visual examination, or by dynamic cone methods to enable measurement of vertical compaction consistency and strength. A typical maximum DCP cone penetration index would be 15mm per blow at an elapsed time of 72 hours after final compaction (ref para 4.3 in Appendix 1).

- Stiffness Modulus by use of the the light dynamic plate to determine modulus (i) immediately upon installation – minimum 20MPa, (ii) after an elapsed time period of 48 hours – minimum 50MPa and (iii) after a further 48 hours an increase of 20% over the 48 hours value. This time related measurement process confirms the onset of strength gain through hydration and maturing. The maximum limit of the measurement range of the light dynamic plate may be reached before the 96 hours elapsed time period but this will be due to the quantity and nature of the hydraulic binders which have been employed and illustrates the intended objective of achieving minimum strengths and strength gains in early life (ref para 4.4 in Appendix 1).

- Volume Stability based upon a volumetric swell criteria of 2.5%. There is no confinement of the specimen which is maintained in contact with circulating, aerated water at 20°C for 90 days. This is a safeguard against the potential presence of sulfates which could lead to expansion. In practice, the period of test precludes this as a control test during construction and much of the information needed to evaluate the prospective risk of volumetric swell will have been undertaken at the design and mixture characterisation stage of the process (ref para 4.6 in Appendix 1).

- Permeability described by means of an accelerated test method conducted upon undisturbed specimens extracted from the as-installed works. The purpose of this requirement is to confirm the data which will have been generated by the mixture characterisation phase and to provide surety to prevent the ingress of water into the hydraulic bound foundation (ref para 4.7 in Appendix 1). The permeability value is generally readily achieved by well compacted medium to high plasticity clays within the acceptability criteria when combined with hydraulic binders of a minimum total mass of 8%.

Full scale construction

The principle of adopting HBFM in lieu of granular sub base was selected early in the design of A428 Crick By-Pass for reasons of construction sustainability. The specification style described in Appendix I was used in the procurement of the works contract with construction works undertaken in 2001/02. The clay soil on the site was largely Lower Lias clay of Jurassic age and had been investigated both in the field and the laboratory using the processes described previously. Since much of the vertical alignment was in cutting the use of HBFM obviated the need for further foundation excavation, disposal of the material arising and importation of granular sub base. HBFM was therefore specified for use throughout the works although for programming reasons some sections of foundation were constructed with granular sub base. The HBFM was constructed in two layers with a total thickness of 350mm using lime and cement binders.

Plate 2, HBFM pulverisation and stabilisation processing

Typical specimen data measured during construction works on the HBFM is illustrated in Figures 2, 3 and 4 with MCV apparatus and an HBFM specimen shown in Plate 3.

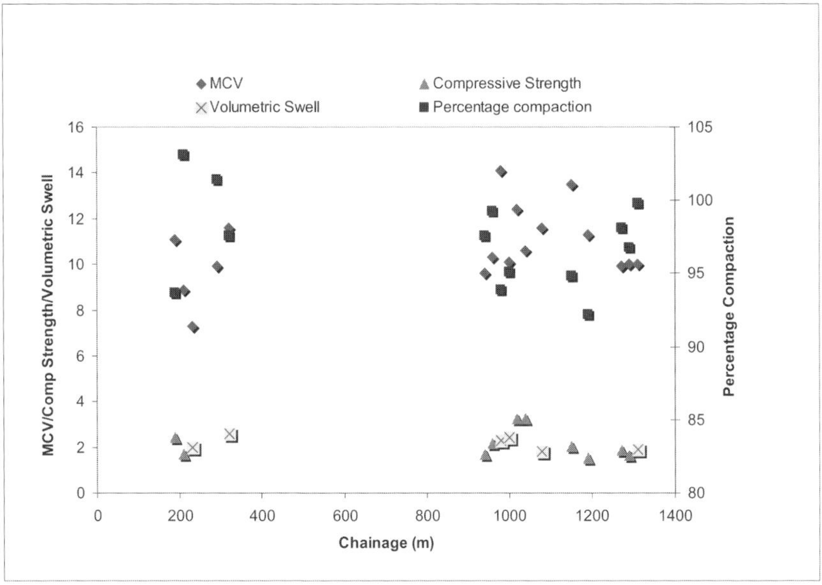

Figure 2, MCV Compressive strength, volumetric swell and compaction data

(a) (b)

Plate 3, Specimen preparation and testing during construction ; (a) Moisture Condition Value apparatus, and (b) HBFM compacted specimen

The Lias clay which forms the parent subgrade soil type at Crick is intrinsically stiff in nature yielding high MCV values and careful control of pulverisation and added water content was necessary to ensure that the material was within the acceptable range for compaction in the works. In general, the percentage compaction, as compared to MCV refusal density, was in most cases 94% or better indicating a high degree of as-installed compaction in the works as illustrated in Figure 2. The relationship between compressive strength and time is shown in Figure 3 where the percentage gain in strength is expressed as

The by-pass road was opened to public traffic in May 2003 and the road pavement will be monitored by rolling wheel transient loading deflection techniques using a Deflectograph machine on a periodic basis to evaluate and analyse the long term HBFM performance in-service.

Acknowledgements

The author is indebted to many colleagues in industry who have contributed, knowingly or otherwise, in the content of this paper and in particular to John Kennedy for inspiration in respect of the A421 Tingewick scheme and to Mike Bordiss, County Highways Officer, Northamptonshire CC for agreement to include the A428 Crick By-pass data.

References.

Rogers C.D.F., Glendinning S., and Dixon N., (1996), (eds). *Lime Stabilisation* Thomas Telford, London. ISBN 0-7277-2563-7

Higgins D. D. and Kennedy J (1999) *Lime and Ground Granulated Blast Furnace Slag Stabilisation of Boulder Clay on the A421 Tingewick By-Pass.* Proc 3rd European Symposium on the Performance and Durability of Bituminous Materials and Hydraulic Stabilised Composites; University of Leeds,

Chaddock B.C.J., and Atkinson V.M (1997) *Stabilised Sub-bases in Road Foundations - Structural Assessments and Benefits*, Report 248, TRL, Cowthorne.

Appendix I : Template Specification – Materials Aspects

Hydraulically Bound Foundation Material (HBFM)

1. General

1.1. This clause shall apply to the production and placement of hydraulically bound material for use as road foundation.

1.2. The material to be stabilised with hydraulic binders shall be Class 7I selected cohesive material complying with Clause 601 and Table 6/1.

1.3. Hydraulic binders shall be:

i). Quicklime of hydrated lime complying with BS890. Quicklime shall when sieve have 100% passing a BS 10mm sieve and at least 95% by mass passing a BS 5mm sieve.
and, either

ii). Ground granulated blastfurnace slag complying with BS6699: 1992

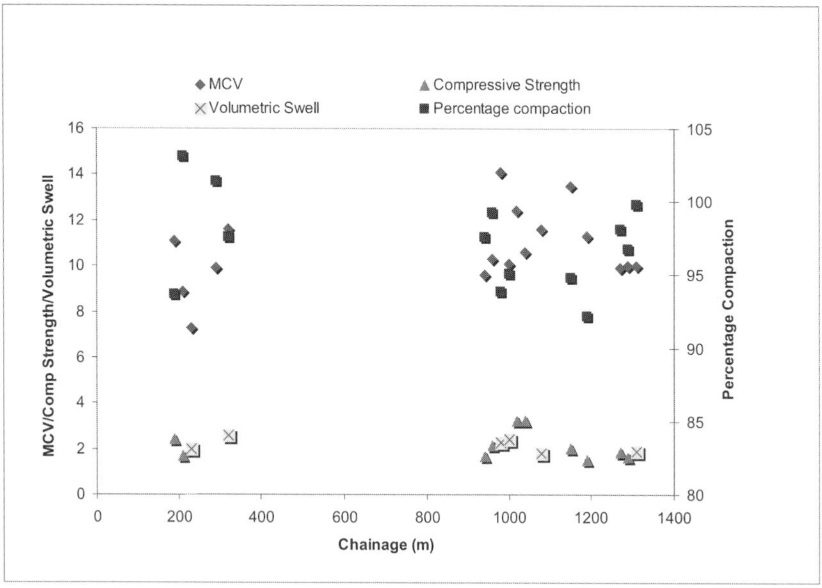

Figure 2, MCV Compressive strength, volumetric swell and compaction data

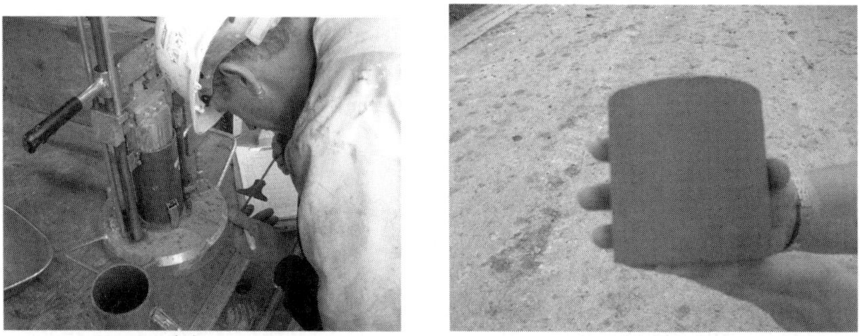

(a) (b)

Plate 3, Specimen preparation and testing during construction ; (a) Moisture Condition Value apparatus, and (b) HBFM compacted specimen

The Lias clay which forms the parent subgrade soil type at Crick is intrinsically stiff in nature yielding high MCV values and careful control of pulverisation and added water content was necessary to ensure that the material was within the acceptable range for compaction in the works. In general, the percentage compaction, as compared to MCV refusal density, was in most cases 94% or better indicating a high degree of as-installed compaction in the works as illustrated in Figure 2. The relationship between compressive strength and time is shown in Figure 3 where the percentage gain in strength is expressed as a proportion of the measured 28 day strength. After fairly rapid hydration in the first 14 days the strength gain gradually increases reaching a `plateau` zone after about 60 days.

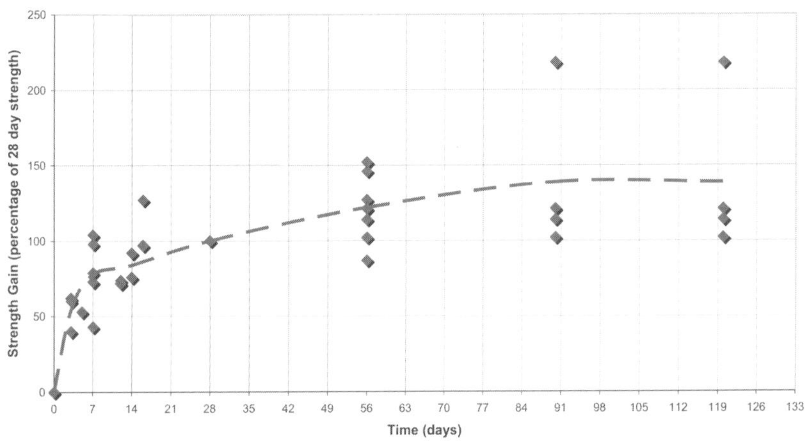

Figure 3, Percentage gain in HBFM compressive strength with time

The light dynamic plate and cone penetrometer test devices, as shown in Plate 3, were used in both layers of the HBFM construction works to confirm that design requirements had been achieved and that the maturing process was proceeding in the field as anticipated and predicted from the laboratory mixture characterisations carried out previously. Both of these items of test equipment yield immediate results which can usefully be employed in monitoring and control of the HBFM works. Natural variations in the soil and construction operations associated with the in-situ works are reflected in the measurements gleaned from test devices such as these and an overall assessment needs to be undertaken in the circumstances which reflect these variables rather than on an individual test result basis. Typical test data from the Light Dynamic Plate is illustrated in Figure 4 which shows an increasing stiffness modulus with elapsed time for a period up to 96 hours after placement and final compaction of the HBFM confirming that the time related maturing process is proceeding as anticipated.

(a) (b)

Plate 4, In-situ testing during HBFM construction with (a) Light Dynamic Plate, and (b) Cone Penetrometer

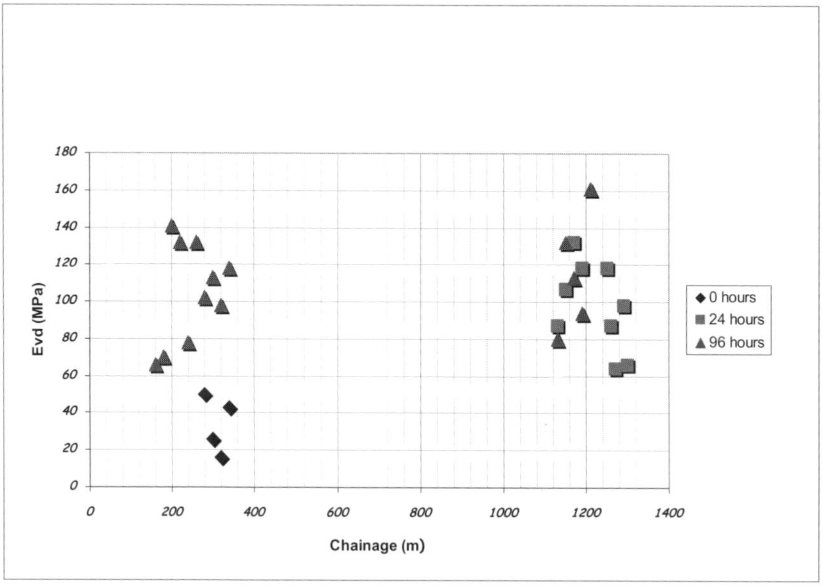

Figure 4, In-situ early life stiffness modulus increase with time

The upper layer of HBFM was sealed with a tack coat curing spray and the overlying hot-mix asphalt placed shortly thereafter. The asphalt base was placed and compacted with conventional plant in the usual way. Specified density requirements in the asphalt material were achieved without any change in the normal compaction methodology thereby illustrating that the HBFM layer had sufficient resilience to provide an adequate compaction platform for the overlying road pavement layers.

The by-pass road was opened to public traffic in May 2003 and the road pavement will be monitored by rolling wheel transient loading deflection techniques using a Deflectograph machine on a periodic basis to evaluate and analyse the long term HBFM performance in-service.

Acknowledgements

The author is indebted to many colleagues in industry who have contributed, knowingly or otherwise, in the content of this paper and in particular to John Kennedy for inspiration in respect of the A421 Tingewick scheme and to Mike Bordiss, County Highways Officer, Northamptonshire CC for agreement to include the A428 Crick By-pass data.

References.

Rogers C.D.F., Glendinning S., and Dixon N., (1996), (eds). *Lime Stabilisation* Thomas Telford, London. ISBN 0-7277-2563-7

Higgins D. D. and Kennedy J (1999) *Lime and Ground Granulated Blast Furnace Slag Stabilisation of Boulder Clay on the A421 Tingewick By-Pass.* Proc 3rd European Symposium on the Performance and Durability of Bituminous Materials and Hydraulic Stabilised Composites; University of Leeds,

Chaddock B.C.J., and Atkinson V.M (1997) *Stabilised Sub-bases in Road Foundations - Structural Assessments and Benefits*, Report 248, TRL, Cowthorne.

Appendix I : Template Specification – Materials Aspects

Hydraulically Bound Foundation Material (HBFM)

1. General

1.1. This clause shall apply to the production and placement of hydraulically bound material for use as road foundation.

1.2. The material to be stabilised with hydraulic binders shall be Class 7I selected cohesive material complying with Clause 601 and Table 6/1.

1.3. Hydraulic binders shall be:

- i). Quicklime of hydrated lime complying with BS890. Quicklime shall when sieve have 100% passing a BS 10mm sieve and at least 95% by mass passing a BS 5mm sieve.
  and, either
- ii). Ground granulated blastfurnace slag complying with BS6699: 1992

or
iii). Portland cement complying with Clause 1001
as described in Appendix 7/1

1.4. The Contractor shall make available to the Engineer technical analyses of each of the hydraulic binders and in the case of lime must include a chemical analysis for 'available lime'. Reports shall be submitted to the Engineer prior to the incorporation in the permanent works and at weekly intervals during periods when hydraulic binders are being used.

1.5. Hydraulic binders shall be added to the processed Clause 7I material (selected cohesive material) to produce a uniform material with the addition of water as necessary. Two differing hydraulic binders shall be added at successive stages in the process in the approximate ratio 1:3 The first binder additive shall be calcium oxide to produce uniformity of the natural soil and to provide high alkalinity. The second binder additive shall be ground granulated blast furnace slag or ordinary Portland cement to produce cementitious properties by activation in the alkali environment. The minimum total hydraulic binder content shall be not less than 8% by mass.

2. Demonstration Trial

2.1. The Contractor shall undertake a demonstration trial of an area of at least 400 square metres prior to commencement of the main works. In the trial the plant, machinery and material process shall be the same as proposed for the main works.

2.2. The trial area may be accepted into the permanent works subject to the agreement of the Engineer.

3. Job Standard Mix

3.1. The Contractor shall declare a Job Standard Mix (JSM) prior to commencement of the permanent works setting out details of materials and hydraulic binders, added water content, stiffnesses and durability, etc, together with compaction plant, layer thicknesses and number of passages of compaction plant. The JSM shall be subject to approval by the Engineer.

4. Material, Placement and Validation

4.1. The maximum compacted thickness of any one layer shall be 200mm. If satisfactory performance is demonstrated in the trial area the layer thickness may be increased to 250mm subject to the agreement of the Engineer.

4.2. Hydraulically Bound Foundation Material (HBFM) shall have a Moisture Condition Value as determined by BS 1924: Part 2 Section 2.2, in the range described in Appendix 7/1 immediately prior to final compaction and incorporation into the works. The material shall be compacted to a wet density of not less than 90% of refusal density as determined by the MCV when tested in accordance with BS1377: Part 9 Test Method 2.4. Specimens for determination of wet density shall be taken within six hours of completion of final compaction.

4.3. The thickness of each layer of compacted HBFM shall be determined by either dynamic cone penetrometer test as described in Clause ABC or by hand excavated trial pits and string line every 250 linear meters as described in Appendix 7/1. Each layer shall display uniform strength throughout the full depth of the layer and, where required, a maximum Cone Penetration Index at an elapsed time of 72 hours (± 4 hours) after final compaction as described in Appendix 7/1.

4.4. The stiffness modulus of HBFM shall be determined on each layer by the dynamic plate test described in Clause LMN. Immediately after installation the stiffness modulus shall be not less than 20MPa. After an elapsed time of 48 hours (± 4 hours) from the final compaction the stiffness modulus shall not be less than 50 MPa. In the following 48 hours, ie 96 hours (± 4 hours) after installation, the value of stiffness modulus shall display an increase of not less than 20% over the 48 hours values.

4.5. The compressive strength of cylindrical specimens manufactured from HBFM using MCV compaction at the time of installation shall be not less than 1.0 MPa. The specimens shall be sealed and cured as described in BS1924 Part 2 Section 4.1.6. (Note: The cylindrical MCV specimen will not comply with the dimensional requirements of BS1924).

4.6. The volume of stability of HBFM shall be determined on cylindrical specimens manufactured by MCV compaction at time of installation. The specimens shall be sealed and cured in accordance with BS1924 for the initial period and thereafter maintained in circulating, aerated water at 20°C (± 2hours). After a total period of 90 days the specimens shall display a volumetric swell of less than 2.5%.

4.7. HBFM materials shall display a permeability of not less than 1×10^{-9} metres per second when sampled in accordance with BS1377: Part 9 Section 2.4, sealed and air cured for seven days and tested in accordance with Clause XYZ. Samples for determination of permeability shall be taken within six hours of completion of final compaction.

Clause ABC – Dynamic Cone Penetrometer Test

1. Dynamic Cone Penetrometer Tests shall be carried out using a device incorporating an 8kg steel drop weight falling vertically through 576mm and making contact with a relatively light steel anvil.

2. The anvil shall be rigidly attached, via steel roads (less than 20mm diameter), to a 20mm diameter 60° cone which is driven vertically downwards by repeated blows from the drop weight falling through its full height.

3. A suitable, independent, means of recording penetration of the cone in millimeteres shall be provided.

4. The Cone Penetration Index, CPI, (blows per millimetre) shall be recorded.

5. Where the Dynamic Cone Penetrometer is being used to determine, indirectly, parameters other than CPI eg., CBR or Modulus, then published relationships should be used to derive these data. The source of this correlation shall be stated.

Clause LMN – Light Dynamic Plate Test

1. Light Dynamic Plate Tests shall be carried out using equipment which has been properly calibrated to the manufacturer's specification and subject to a validation check prior to use.

2. The equipment shall be capable of delivering a total load pulse of peak magnitude 6-8kN, of total duration 15-40 milliseconds, to a rigid circular plate of 300mm diameter. Both the applied load and the transient deflection shall be measured.

3. The dynamic stiffness modulus shall be determined at each point tested using the following formula:

$$\text{Dynamic Modulus, Evd (MPa)} = \frac{P(1-x^2)}{0.3y}$$

Where: P is the peak applied load (kN)
y is the peak deflection (mm)
x is the Poissons Ratio (a value of 0.35 shall be used in the absence of any other data).

4. The full technical specification for the Dynamic Plate Test apparatus is published by the German Federal Ministry of Transport, Road Construction Department in TP BF-StB Part B 8.3, 1992 (In German).

Clause XYZ – Method for Accelerated Triaxial Permeability Test

1. The accelerated permeability test method is based on procedures for the determination of the permeability of cohesive materials described in BS1377: Part 6: 1990 with the following modifications.

2. The accelerated method is to prepare and set up the sample in the triaxial cell as described in BS1377 and immediately start permeating the sample by applying pressure difference between the top and bottom of the sample in addition to the cell confining pressure. Depending on the initial condition of the sample the sample will usually take in water at both ends or will consolidate from both ends until it has reached a stable condition.

3. As the sample stabilises, permeation starts to occur. The quantity of water passing into the base of the sample and out of the top are measured independently until there is linear flow through the sample. This is calculated as rate of flow in ml/minute and is used to determine the permeability of the specimen as described in BS1377.

4. At this stage, if required the degree of saturation can be determined by measuring the pore water pressure response with an increase in the cell confining pressure. If the resulting "B" value is not greater than 0.95, permeation is achieved. It has generally been found that samples are adequately saturated by the first check.

Bioengineering and the transportation infrastructure

J.R. Greenwood, J.E. Norris, J. Wint, and D.H. Barker
Nottingham Trent University

Introduction

Vegetation will generally establish itself naturally over time even on relatively barren soils provided some nutrients and water are available. In the United Kingdom most soil slopes will support an array of vegetation types. Grasses, shrubs and trees will initially self-seed as 'pioneer' vegetation and evolve into a consistent pattern of coverage referred to as 'climax' vegetation.

Embankment and cutting slopes formed as part of the UK transportation infrastructure are generally seeded with grasses in accordance with the Specification for Highway Works (Design Manual for Roads and Bridges and Manual of Contract Documents for Highway Works) and selected shrubs and trees planted in accordance with locally agreed landscaping criteria. Whilst it is recognised that the grass, once established, will prevent surface erosion, the vegetation is not intended for any purpose other than landscaping aesthetics.

Many of the embankments and cutting slopes in the UK, particularly in the South East of England, are constructed of or within stiff over-consolidated clays which are prone to softening with time leading to shallow slope failures (Greenwood et al, 1985). It is becoming increasingly important, as the need for more eco-friendly solutions arises, for engineers to explore how vegetation might be selected and maintained, to help enhance the soil strength and reduce the risk of shallow slope failure.

However, the detrimental effects of vegetation cannot be ignored. Figure 1 indicates some of the problems frequently encountered due to vegetation when it exists in the 'wrong' locations in relation to engineering constructions. The detrimental effects on foundations located too close to certain trees leading to ground movements of a seasonal and permanent nature has been extensively studied by the Building Research Establishment (1987) and others (Biddle, 1998).

On the other hand, vegetation can often be seen 'holding together' slopes that would otherwise degrade very rapidly. Examples are shown in Figures 2. There is a general awareness and perception by the public that tree roots bind the soil

a). Damage to pavements due to tree roots, Nottingham Trent University car park

b). Retaining wall damage due to roots at Nottingham Trent University

c). Wedging apart of sandstone blocks due to roots in fissures at Nottingham Castle

Figure 1, Examples of detrimental effects of vegetation

a) Water Lane, Kent. Vegetation root network permits steeper slopes in Greensands

b) and c) Dune grasses resisting erosion and local instability on the Wash

Figure 2, Examples of vegetation assisting stability

together as indicated by the following report from the Daily Telegraph following a minor train derailment near Merstham tunnel on the Brighton line on 2nd January 2003:-

'Passengers rescued as train hits mudslide'.
...Sam Livermore, whose home is beside the track, said: "Since they uprooted trees about 10 months ago the banks have become increasingly unstable as there are no longer any roots to keep the ground in place."
A resident whose home overlooks the cutting said: "Last year work was carried out to supposedly prevent land-slips. But workers missed a 40 yard section when they were putting wire netting and reinforcing materials on the bank. It is this exact spot where the landslip has happened. It is on a bend so the train driver would have been on top of it before he realised."
A spokesman for Network Rail said the trains had been ordered to travel at 5 mph because of the heavy rainfall over the previous 24 hours. He said almost an inch of rain had fallen during that time. "The trees were taken out because of the risk of them falling on to the tracks," he added. "They presented more of a risk than landslides and contrary to popular belief they do not make the embankments more stable."

It is interesting to note that the Network Rail 'spokesman' was rather dogmatic that the trees were not helping stability but the danger of them falling on to the track was high. The contribution and problems associated with presence of vegetation on the London Underground cutting and embankment slopes was more positively discussed by Gellatley et al, (1995). There is an obvious need to quantify the potential benefits (and dis-benefits) that vegetation can bring to the stability of slopes.

This paper summarises the work relating to Soil Bioengineering carried out at Nottingham Trent University and assesses the various influences that vegetation will have on the stability of slopes.

CIRIA vegetation trials

The publication by CIRIA of the text 'Use of Vegetation in Civil Engineering' (Coppin and Richards, 1990) formed a major landmark in introducing to engineers the concepts of enhancing soil properties with appropriate vegetation. This was followed up by further guidance relating specifically to highway slopes (Barker, 1996, 1997) and CIRIA sponsored field trials of specific vegetation on the M20 motorway at Longham Wood, near Maidstone, Kent (Greenwood et al, 2001).

The main influences of vegetation are given in Figure 3, based on Coppin and Richards (1990). The M20 trials set out to assess the relative importance of these influences on the geotechnical parameters and stability of a slope (Greenwood et al, 2001). The Longham Wood site was monitored for a period of 5 years after which it had to be destroyed as the new Channel Tunnel Rail link was constructed immediately adjacent to the M20 passing through the trial site.

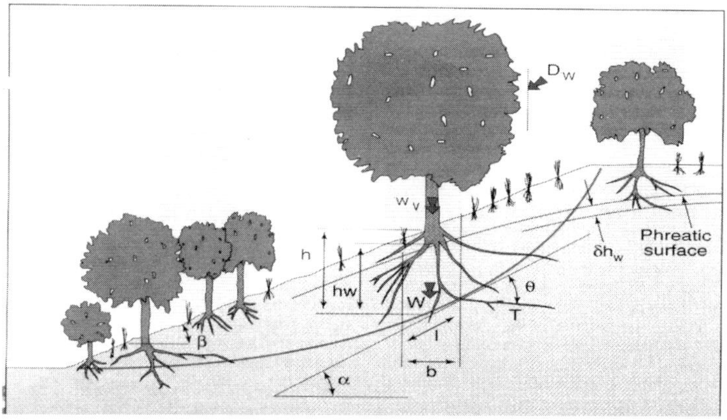

Basic parameters and dimensions used in stability analysis by method of slices

Term	Units	Description
H	m	Average height of slice
B	m	Width of slice
ℓ	m	Length (chord) along base of slice
c'	kN/m^2	Effective cohesion at base of slice
ϕ'	degrees	Effective angle of friction at base of slice
γ	kN/m^3	Bulk Unit weight of soil in slice
γ_w	kN/m^3	Unit weight of water (usually taken as 10 kN/m^3)
W	kN	Total weight of soil in slice (for layered soils, 1,2,3 etc $W = (\gamma_1 h_1+\gamma_2 h_2+\gamma_3 h_3+\text{etc}) \times b$)
α	degrees	Inclination of base of soil slice to horizontal (may be negative at toe)
h_{w1}	m	Height of free water surface at left hand side of slice
h_{w2}	m	Height of free water surface at right hand side of slice
U_1	kN	Water force on left hand side of slice (from flow net, seepage calcs or based on h_{w1})
U_2	kN	Water force on right hand side of slice (from flow net, seepage calcs or based on h_{w2})
h_w	m	Average piezometric head at the base of the slice. For hydrostatic $h_w = (h_{w1} + h_{w2})/2$
U	kN/m^2	Average water pressure on base of slice ($= \gamma_w \times h_w$)
F	ratio	Factor of Safety (usually shear strength/ shear force on slip plane)
F_m	ratio	Factor of Safety in terms of moment equilibrium
F_f	ratio	Facto of safety in terms of horizontal force equilibrium
Vegetation, Reinforcement and Hydrological effects		
c'_v	kN/m^2	Additional effective cohesion at base of slice (due to vegetation etc.)
W_v	kN	Increase in weight of slice due to vegetation (or surcharge)
T	kN	Tensile root or reinforcement force on slice
θ	degrees	Angle between direction of T and base of slip surface
Dw	kN	Windthrow force (downslope)
β	degrees	Angle between wind direction and horizontal (often assume equal to slope angle)
Δh_{w1}	m	Increase in height of free water surface at left side of slice
Δh_{w2}	m	Increase in height of free water surface at right side of slice
ΔU_1	kN	Increase in water force on left hand side of slice
ΔU_2	kN	Increase in water force on right hand side of slice
Δh_w	m	Increase in average piezometric head at base of slice (due to vegetation)
Δu_v	kN/m^2	Increase in average water pressure at the base of the slice, $= \gamma_w \times \Delta h_w$

Figure 3, The various influences of vegetation (developed from Coppin and Richards, 1990) and notation used for routine stability analysis by the method of slices (Greenwood, 1989).

During the final 'destructive' testing of the site, trenches were excavated to provide more detail of the ground and root growth conditions (Figures 4a–4b). Apparatus was developed to assess the in situ shear strength of the root reinforced Gault Clay and to determine the resistance of selected roots to pulling out of the ground.

The main conclusions from the Longham Wood trials were (Greenwood et al, 2001):-

- Willow and alder trees became established over the five year trial period and developed a substantial root network extending to 1.2 m depth.
- The instrumentation used, particularly that for determining soil water pressure, detected changes in the state of the slope produced by the vegetation and root systems.
- Seasonal changes in ground conditions were clearly indicated by the Mackintosh probe testing but this testing was not sensitive to the smaller changes due to the vegetation.
- The counterfort slope drains had no apparent effect on the vegetation or the soil and groundwater conditions in the upper 1.2 m of the slope.
- Of the possible influences of the vegetation, the tensile root force was found to be most effective in increasing the resistance to slope failure.
- The study recommended that further monitoring is carried out on other sites to examine the effects of the vegetation in the medium - long term and to quantify the strength contribution available from different root systems.

Moisture content changes during the trials were monitored by use of a neutron probe inserted down access tubes at specific locations (Vickers and Morgan, 1999). During the final 'destructive' testing physical moisture contents were taken and the 'moisture in the bag' technique Greenwood and Norris (1999a).

a) Final trench with roots present to 1.2m

b) Roots concentrated around the Neutron Probe access tube

Figure 4, Trenching at the end of the M20 trials to check on root growth with depth.

The ECOSLOPES 5th framework project

The award of a £1.6m research grant under the 5th Framework of the European Community enabled Nottingham Trent University, as a partner in the ECO-SLOPES project, to further develop the in situ shear apparatus and to link the work done in the UK with related work in other European countries. The project is broad-based with the partners focusing on the many related aspects of vegetation as listed in Table 1. Current details of the project are available on the Website, www.ecoslopes.com.

It is intended that the final outcome of the project will be a reference data base and a manual or computer-aided decision support system to help the slope engineer to select, specify and maintain appropriate vegetation to enhance slope stability in the various regions of Europe.

Table 1, The EU funded ECOSLOPE partners and research activities (Contract period 2001–2004)

ECOSLOPES PARTNER	PARTICULAR ACTIVITY
NTU (Nottingham Trent University)	Root investigations -shear and pull out testing; Stability analysis; Vegetated slope data base; Decision support system.
INRA (University of Bordeaux)	Project coordination; Root architecture; Tree winching; Numerical modelling (with Wilde and Partners).
Cemagref (France)	Dynamic effects – vegetation to resist rockfalls.
Forest Research, Scotland	Forest Stand stability.
University of Molise, Italy	Root architecture.
Geostructures, UK (joined NTU)	Modelling; Decision support system.
NAGREF, Forest Research, Greece	Effects of fires on vegetation and erosion.
IBED, University of Amsterdam	Site characterisation; Modelling.
CIDE, Spain	Desertification; Forest fires; Vegetation recovery.
End User Group (UK reps Alex Kidd, Neil Bayfield)	Comments and guidance to research contractors.

The influences of vegetation (and how they may be modelled)

In this section each of the possible influences of vegetation on a slope (Figure 3) is reviewed in the light of the M20 trials, the ECOSLOPES project and reference to other work.

Enhanced cohesion, c'_v.

The concept of effective cohesion in soils has received considerable attention with some researchers advocating that no true cohesion exists in clay soils. However back analysis of slope failures has generally indicated an operational effective shear strength which is best represented by a small cohesion intercept in the order of $c' = 1$ to $2\ kN/m^2$. The actual value can have considerable influence on the calculated factor of safety, F, hence the interest of geotechnical engineers in defining the value.

It would be expected that a fine root network would act to provide an enhanced cohesion much in the same way that geosynthetic mesh elements have been demonstrated to enhance the soil strength properties (Andrawes et al, 1996).

Values of c'_v have been measured by researchers often based on shear tests (Coppin and Richards, 1990, Table 3.4).

The use of enhanced c′ values will be appropriate for grassed areas or areas of uniform vegetation where fine root distribution with depth is consistent and easily defined (see later and Figure 6).

In general the reliable benefit of an enhanced c′ value will be limited to shallow depths. Just as it is difficult to measure accurate values of c′ which are appropriate for stability analysis it will be equally difficult to measure the additional contribution, c'_v, due to the vegetation. Field tests will tend to give an indicative undrained strength increase due to the presence of fine roots but, for clay soils, the true effective parameters are probably best obtained by back analysis or more sophisticated effective stress laboratory testing.

The role of fine roots in resisting surface erosion is well documented (Morgan and Rickson, 1995). Whilst fine roots are the major root components in garnering nutrients and moisture from the soil, their role in more general slope stability is less certain with perhaps a minor contribution as they help to maintain the integrity of the surface layers.

The Mass of Vegetation, W_v.

The mass of vegetation is only likely to have a major influence on slope stability when larger trees are present. The loading due to a well stocked forest of 30 to 50m tree height is in the order of 0.5 to 2 kN/m^2 (Coppin and Richards, 1990, Figure 3.17). A 30m high tree having a base trunk diameter of around 0.8m is likely to have a weight of around 100 to 150 kN. Such trees located at the toe of a potential slip could add 10% to the factor of safety. (See Coppin and Richards, 1990, Figure 3.18). Equally if located at the top of a potential slip the factor of safety could be reduced by 10%. Each situation must be individually assessed for the mass of vegetation involved. It should be borne in mind that plant evapo-transpiration will reduce the weight of soil as moisture is lost. This can be important on slopes of marginal stability.

When larger trees are removed from the toe area of a slope, in addition to the gradual reduction in soil strength due to the loss of evapotranspiration effects, the reduction in applied loading could result in temporary suctions in clay soils which may lead to softening as available water is drawn in to satisfy the suction forces. This is of course akin to the recognised softening of overconsolidated clays due to relaxation of overburden pressures when placed in the top layers of an embankment from deep cutting (Greenwood et al, 1985).

Windthrow loading, D_w.

Windthrow loading is particularly relevant when considering the stability of individual trees but of lesser significance for general slope stability where the wind forces involved represent a much smaller proportion of the potential disturbing forces and trees within a group (stand) are sheltered to some extent by those at the edge.

Windthrow forces on single trees may be estimated from Brown and Sheu (1975), and windthrow on forested slopes may be calculated (Hsi and Nath, 1970). (Both approaches given in Coppin and Richards, 1990).

Soil strength increase due to moisture removal by roots.

There have been various well documented observations of moisture deficit around trees (Biddle, 1998) due to the effects of evapotranspiration and the problems this has caused for buildings (Hunt et al, 1991). However when it comes to relying on tree and shrub roots to remove water and hence strengthen the soil it is not quite so straightforward.

Observations on the M20 at the Longham Wood trial site indicated huge seasonal variation in the moisture content (and hence the undrained soil strength) of the south facing trial area (Figure 5). These large variations masked any effects due to the vegetation over the 5 year period of the trials (Greenwood et al, 2001).

More work is needed to compare the moisture contents of slopes with particular types of vegetation with adjacent slopes in the same soil type without vegetation (or with grass alone). The availability of Time Domain Reflectometry and Theta probe technology to assist in non destructive moisture content determinations should enable data to be accumulated on the actual influences of the vegetation on moisture content.

During particularly wet periods, the ability of the roots to influence the seasonal moisture content will be curtailed and therefore any enhanced soil strength gained previously by evapotranspiration will be reduced or lost entirely to an extent difficult to quantify. Hence this effect cannot be taken into account at such critical times. However, it can be assumed that there is a narrowing of the window of risk of failure due to soil saturation by storm events or periods of prolonged rainfall. Furthermore, whilst moisture content changes influence the undrained shear strength (c_u) the effective stress parameters ($c'$and ϕ') as generally used in routine stability analysis are not directly influenced by the changing moisture content, although the water pressures (suctions) used in the analysis may well be.

It should be borne in mind that desiccation cracks, possibly extended during dry periods by the presence of certain vegetation, will encourage a deeper penetration of water and water pressures into the soil during wet periods. However, these cracks will subsequently provide pathways for roots to extend deeper into the soil in their search for moisture and nutrients.

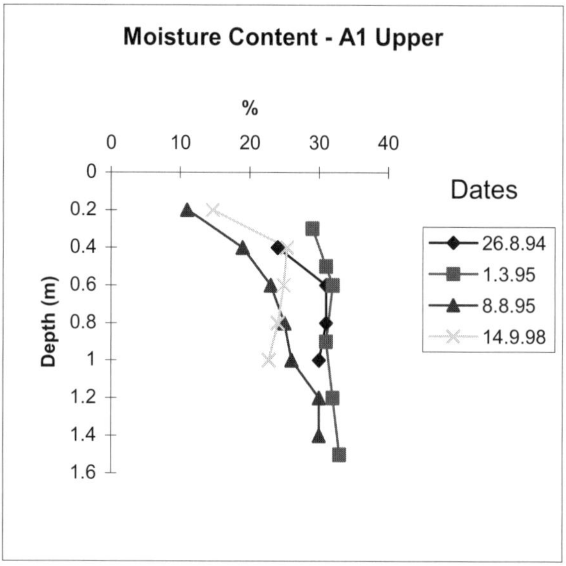

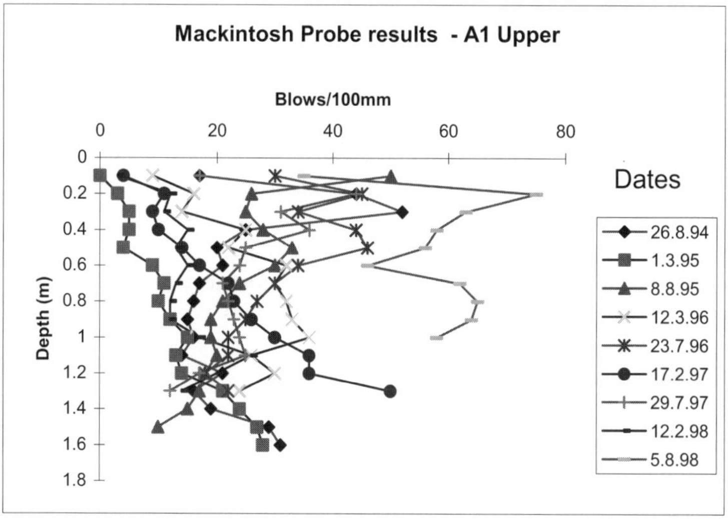

Figure 5, Typical moisture content and Mackintosh probe results from M20 vegetation trial site indicating extreme seasonal variations in moisture content and soil strength (Greenwood and Norris, 1999b).

Suctions and changes in pore water pressure due to vegetation (u_v).

As discussed in the previous section, the moisture content and soil water pressures are related. On the M20, seasonal fluctuations in the water table, as measured by standpipes, were not significantly modified by the effects of the newly established vegetation. Tensiometers installed on the M20 project (Vickers and Morgan, 1999) and on other slopes have proved much more worthwhile in recording the detailed response of the ground suctions to rainfall events and periods of wet or dry weather.

The soil scientist and agriculturalist has tended to view soil suctions and moisture contents rather differently from the civil and geotechnical engineering approach. It is recognised that there is some merit in relating the geotechnical engineering parameters to the terminology of the soil scientist. Some terms which are relevant to the consideration of the effects of vegetation are described below with their relationship to conventional geotechnical terms.

- *Soil moisture characteristic curve* – this relationship between the moisture content and the suction pressures is particularly relevant to the geotechnical engineer (Fredlund and Xing, 1984).
- *Field Capacity* – the moisture content after saturation and after free drainage has practically ceased. Typical suctions are -5 to -10 kNm^2 at field capacity.
- *Moisture deficiency* – the difference between the measured moisture content and the field capacity.
- *Gravimetric (engineering) moisture content* = mass of water / mass of solids (dry soil).
- *Volumetric moisture content* (as used by soil scientists and measured by indirect tests such as Theta probe) = Volume of water / Total volume of soil.
- The gravimetric and volumetric moisture contents are related by:-
- Gravimetric moisture content = Volumetric moisture content x density of water / dry density of the soil.

Tensile root strength contribution, T.

The tensile strengths of roots of various diameters from different species have been measured in the laboratory and found to be typically in the order of 5 - 60 MN/m^2 (Coppin and Richards, 1990).

In the field, to make use of the available tensile strength to enhance slope stability the root must have sufficient embedment and adhesion with the soil. The biological growth patterns and interaction between the root and soil are complex (Greenwood et al, 2003 in preparation) but for engineering purposes the available force contribution from the roots may be measured by in situ pull out tests.

Measurement of the root pullout resistance has been carried out by various methods ranging from hand pull to screw and hydraulic jacks. The method depends very much on the size of root and the convenience of available equipment and a reaction frame. A constant rate of strain is required, typically 1% per minute, and a means of measuring the resistance by spring balance or load cell at defined displacements. Procedures for the root pull out test are given in Greenwood et al, (2003 in preparation).

Design of the clamp to grip the root requires particular attention. Many species of root, particularly when fresh, demonstrate a tendency for the bark to separate and slide over the core wood during tensile testing. It is therefore often necessary to strip the bark at the clamp and to grip directly on to the core wood. The tensile strength is then calculated based on the diameter of the core wood assuming that the bark is making little contribution to the strength of the root. However it is the bark which is in contact with the soil and generating the adhesion resistance so the full root diameter must be considered in the pull out assessment. These issues are discussed by Greenwood et al, (2003 in preparation).

Modelling of vegetation influences

The various influences of vegetation on the factor of safety of a slope are conveniently assessed by routine limit equilibrium stability analysis. Various methods of stability analysis are available. The Greenwood General Equation (equation 1) (Greenwood, 1989; Morrison and Greenwood, 1989) is considered particularly appropriate because it takes full account of hydrological (seepage) forces to give a realistic estimate of the factor of safety for all types of slopes and slip surfaces.

$$F = \frac{\sum\left[c'\ell + \left(W\cos\propto - u\ell - \left(U_2 - U_1\right)\sin\propto\right)\tan\phi'\right]}{\sum W\sin\propto} \qquad (1)$$

The mathematically 'simple' form of the equation and the factor of safety defined in terms of restoring and disturbing forces means that it is straightforward to add the various vegetation influences (equation 2)

$$F = \frac{\sum\left[(c'+c'_v)\ell + \left((W+W_v)\cos\propto - (u+\Delta u_v)\ell - \left((U_2+\Delta U_{2v}) - (U_1+\Delta U_{1v})\right)\sin\propto - D_w\sin(\alpha-\beta) + T\sin\theta\right)\tan\phi'\right]}{\sum[(W+W_v)\sin\propto + D_w\cos(\alpha-\beta) - T\cos\theta]} \qquad (2)$$

A procedure for estimating the available tensile root reinforcement force, T, based on observation of the number of roots of a given diameter present at a particular depth is given in Norris and Greenwood (2000). A factor of safety of 8 is applied to the measured pull out resistance to allow for the large strain needed to generate the peak measured root pull out force and for other uncertain factors relating to root distribution.

An EXCEL spreadsheet, known as 'SLIP4EX', has been developed at Nottingham Trent University to compare routine methods of analysis for a given slip surface and to quantify the changes to the factor of safety due to the influences of the vegetation.

The tensile root force available and other changed parameters due to the vegetation may be assessed by considering the typical distribution of roots below a vegetated area.

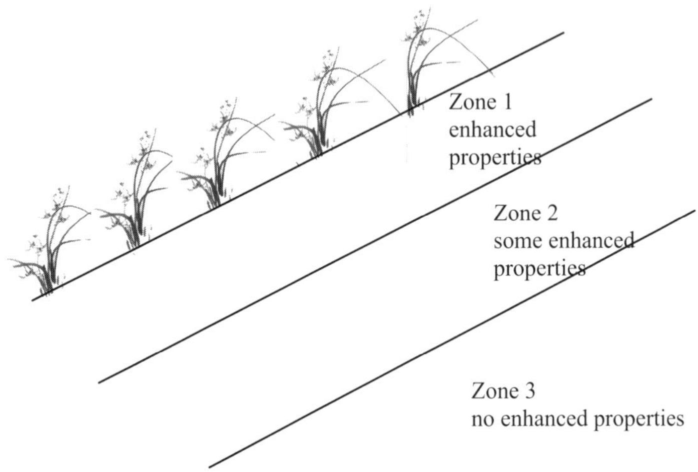

Figure 6, Zones of enhanced soil properties for regular vegetation cover

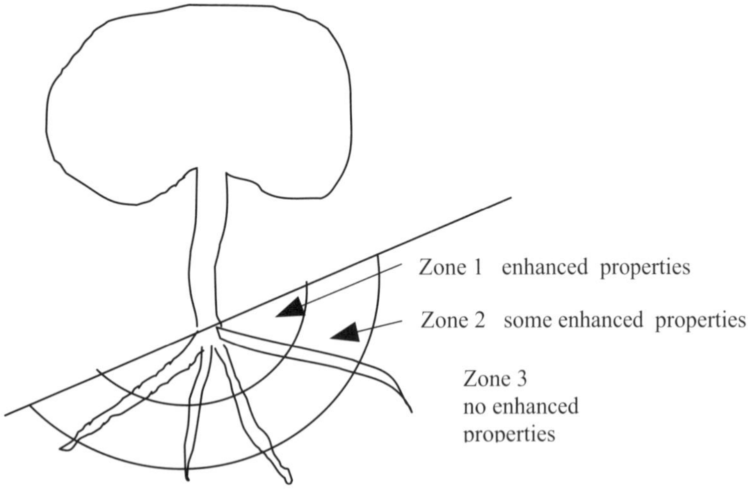

Figure 7, Saucer shaped zones of enhanced parameters beneath a single tree.

If the coverage is consistent over the area, enhanced parameter zones may be represented as zones parallel to the slope (Figure 6). For isolated larger trees and shrubs a distribution such as that shown in Figure 7 might be considered as being typical of the saucer shaped root network frequently encountered.

Finite element modelling of the vegetation influences is also helpful particularly where strain compatibility is to be considered. The application of finite element programs such as PLAXIS to vegetated slopes is being assessed within the ECOSLOPES project.

Conclusions

The presence of vegetation may be sufficient to maintain stability of certain marginally stable slopes. A framework of modelling by limit equilibrium and finite element methods already exists and data are being acquired to help quantify the enhancement that vegetation can provide.

Of the various influences, the physical presence of roots and the tensile reinforcement they can provide appears to be the most significant based on observations to date.

The on–going development of field monitoring and analytical techniques with engineers working alongside the plant specialists, soil scientists and foresters to determine characteristic growth patterns and resulting changes in geotechnical parameters should lead to the necessary guidance on selection and maintenance of the vegetation to assist slope stability.

Future research should address the implications of climate change affecting the long term stability of vegetated slopes. The establishment of a 'controlled climate' test bed on a purpose built embankment of known soil properties will provide the necessary facilities for longer term modelling/monitoring of how vegetation on slopes reacts to changes in climatic conditions.

Acknowledgements

The research team is most grateful for the funding support for this work provided by CIRIA and the 5th Framework of the European Union.

References

Andrawes K.Z., McGown A., Hytiris N., Mercer F.B. and Sweeterland D.B. *The use of mesh elements to alter the stress strain behaviour of granular soils*. Proc. 3rd Int. Conf. on Geotextiles, Vienna, Austria, 1996.

Barker D.H. *Vegetation on highway slopes – establishment*. Highways, Alad Publishing, June, 20-23, London, 1996.

Barker D.H. *Vegetation on highway slopes – maintenance*. Highways, Alad Publishing, April, 24-26, London, 1997.

Biddle, P.G. *Tree root damage to buildings*. Willowmead Publishing, 1998.

Building Research Establishment, *The influence of trees on house foundations in clay soils*. BRE Digest 298, BRE, 1987.

Brown C.B. and Sheu M.S. *Effects of deforestation on slopes.* J. Geotech. Engng Div. ASCE 101, (GT2), Feb, 147-165, 1970.

Coppin N.J. and Richards I.G. *Use of vegetation in Civil Engineering*, CIRIA, Butterworths, 1990.

Design Manual for Roads and Bridges. ISBN 0115524606. 2003.

Gellatley M.J., McGinnity B.T, Barker D.H. and Rankin W.J. *The interaction of vegetation with the LUL surface railway system.* In Proc. Conf. Vegetation and slopes: stabilisation, protection & ecology (Ed. D H Barker), 29/30 September, University Museum, Oxford, Thomas Telford, 60-71, 1995.

Fredlund D. and Xing A. *Equations for the soil-water characteristic curve.* Canadian Geotechnical Journal, Vol. 31, 521-532, 1994.

Greenwood J.R., Holt D.A. and Herrick G.W. - *Shallow slips in highway embankments constructed of overconsolidated clay.* ICE Int. Symp on Earthworks Failures, 6-7 March, 79-92, 1985.

Greenwood J.R. *Effective Stress Stability Analysis.* Discussion in 9th European Conference on Soil mechanics and Foundations, Dublin Sept 1987 Vol 3, Post conference proceedings, Balkema, 1082-1083, 1989.

Greenwood J.R. and Norris J.E. *"Moisture in the Bag", A simplified procedure for the determination of soil moisture content by oven drying.* Ground Engineering, 32, 6, 32-33, 1999a.

Greenwood J.R. and Norris J.E. *Measurement of changes in geotechnical parameters due to the effects of vegetation.* Proc. of the 2nd Int. Conf. on Landslides, Slope Stability and the Safety of Infra–Structures, July 1999, Singapore, 87-94, 1999b.

Greenwood J.R., Vickers A.W., Morgan R.P.C., Coppin N.J. and Norris J.E. Bioengineering - *The Longham Wood Cutting field trial.* CIRIA Project Report 81, London, 2001.

Greenwood J.R., Norris J.E. and Wint J. *A procedure to determine the pull-out resistance of roots.* (in preparation, 2003).

Hunt R., Dyer R.H. and Driscoll R. *Foundation movement and remedial underpinning*, BRE report 184, 1991.

Manual of Contract Documents for Highway Works. ISBN 0115525084. 2003.

Morgan R.P.C. and Rickson R.J. (eds) *Slope stabilization and erosion control: a bioengineering approach.* Spon 1995.

Morrison I.M. and Greenwood J.R. *Assumptions in simplified slope stability analysis by the method of slices.* Geotechnique 39, No 3, 503-509, 1989.

Norris J.E. and Greenwood J.R. *In situ shear and pull out testing to demonstrate the enhanced shear strength of root reinforced soil.* Proc. 8th Int. Symp. on Landslides, Cardiff, June, 1123-1128, 2000.

Vickers A.W. and Morgan R.P.C. *Soil-water monitoring to assess the effectiveness of three bioengineering treatments on an unstable Gault Clay cutting in Southern England.* Proc. Of 2nd Int. Conf. on Landslides, Slope Stability and the Safety of Infra-Structures, Singapore, July, 95-102, 1999.

Legacy of railway construction: impact on geotechnical characterisation

Professor M.S. Rosenbaum and C.F. Taylor
The Nottingham Trent University

Introduction

Derelict railway land is a valuable resource. It is often located near to where people want to live, close to employment and social amenities, with good access and service infrastructure. The characterisation of such land is of particular importance to planning, as well as to development and ownership of land. In the UK, land available for development typically has a long and varied history of previous usage. However, there has been a general failure to recognise that 'derelict railway land' encompasses a wide variety of geotechnical characteristics. The potential re-use of a former railway site will be affected by the historical legacy of its development, the current state of the land, and the presence of hazards caused by past activities.

The value of former railway land

The extent of derelict ground in the UK has been highlighted in recent Government reports concerning land redevelopment and the provision of housing. The White Paper *Planning for the Communities of the Future* (DETR, 1998a) sets a target for at least 60% of new homes to be built on previously used land by 2016. Development pressures have resulted in the need to restrict expansion on floodplains, to make city living desirable once again and to bring brownfield sites back into use. The longer-term aim is to bring all derelict and contaminated land back into beneficial use by 2030 (DETR, 1999): the concept of sustainable urban regeneration. Current UK Government policy reflects its commitment to the sustainability challenge by "...making more efficient use of land by maximising the re-use of previously-developed land, and the conversion and re-use of existing buildings..." (DETR, 2000).

Derelict railway land forms urban corridors which are not only of historical significance but offer considerable potential for regeneration, including future transport facilities for both passenger travel and freight distribution (DETR, 1998b), provided their geotechnical character can be determined and accommodated. But before construction can begin, it is crucial to recognise the potential problems that may be encountered at a specific derelict site.

Transportation geotechnics. Thomas Telford, London, 2003

The characteristics of railway land from a geotechnical perspective

The potential use to which any piece of disused railway land might be put needs to take cognisance of its physical characteristics (Appleton & Appleton, 1970):

- areal extent
- gradient
- curvature
- track bed

There is a considerable variation in the width of railway formations and hence the acreage per length of track. At a width of 20 feet (6.7 metres), one mile (1.6 km) of single track occupies 2.4 acres (1.0 hectare). Some formations may be rather wider than this. Furthermore, cuttings and embankments will take up additional land, and if stations and sidings are included a figure approaching 10 acres per mile (2.5 hectares per kilometre) provides a more realistic estimate.

Stations and their yards also vary greatly in area. Industries have historically been attracted to the railway, and abandonment of adjacent station yards has facilitated extension of industrial sites over the newly released land. A combination of station sites and intervening track produces a kind of 'paternoster' effect, the beads or bulges offering potential for users that have no interest in the track *per se*. Another attraction is the ready access and existing water supply and other services for such sites.

There are no rigid criteria by which gradients and curvature have been set. The major factors are prevailing topography and the importance of the railway line to the network, thus the range is rather wide. Railway lines steeper than 1 in 40 are rare, the steepest mainline in the UK being 1 in 38 (the Lickey Incline). Beyond hilly areas, gradients do not generally exceed 1 in 50. For some categories of use, curvature may be a more limiting factor than gradient. Many rural branch lines, for instance, cannot accord with contemporary motorway standards, although they would be acceptable for many lower classes of road.

The track bed comprises ballast and sub-ballast layers including any geotextiles, geomembranes, meshes or other materials for separating, filtering, waterproofing, draining, reinforcing or other strengthening of those layers. The main functions of the track bed are to support the track, to drain water away from the bottom of the sleepers, and to distribute loads to the subgrade so that the position of the rail returns to its original location after the passage of each train. The ballast forming the track bed may have some monetary scrap value once its use supporting the permanent way is finished but is rarely removed, consequently affecting the ease with which change of land use can be effected. Once the railway has been declared 'disused' or 'derelict', deterioration rapidly sets in, especially as vegetation control measures expire, although the obligation remains to maintain fencing and the safety of structures (especially bridges).

Earthworks

The main earthworks associated with railway land are: cuttings, embankments and the track formation. Their construction and the types of processes each formation is susceptible to will now be considered in turn.

The main objective when excavating a cutting is to balance the volume of material that is to be removed with that required to form the embankments (Williams, 1850). However, knowledge of long term slope behaviour was somewhat limited in the nineteenth century when most were created, and many side slopes have since proved to be too steep, especially those in clay.

Any excess earth to that needed for neighbouring embankments would be "put to spoil", in other words, laid down and spread out on land adjoining the railway. Deficiencies similarly led to cutting virgin ground to produce fresh material – known as "side cutting" (Williams, 1850). This was more cost effective than importing materials from other regions, at least until a comprehensive road and rail network had become established.

Cuttings

During excavation of a cutting, the most vital consideration is groundwater. Springs have frequently been exposed, discharging large quantities of water leading to erosion. Heavy rainfall can make water unexpectedly pour from the cutting sides. Effective drainage systems were therefore necessary, originally of gravel but nowadays of geotextiles or geomembranes to create a drainage layer between the ballast and the subgrade (Cope, 1993). It is interesting to note that cuttings through Chalk were often created with almost vertical walls, not only reducing the volume to be moved but also the material's susceptibility to dissolution and thereby reduce the adverse effects of frost action (Phipps & McGinnity, 2001).

Superficial slips were (and are) common, often causing movement during wet periods whilst construction was underway or in the first or second winter thereafter. Repair usually comprised no more than trimming and installation of a few shallow trench drains.

However, deep slips can occur too, even without warning, several (sometimes many) years after excavation. In 1841, an enormous slip took place on the west side of the 48 ft (~15 m) deep New Cross cutting in brown (i.e. weathered) London Clay, three years after excavation. It had a length of 360 ft (~110 m) and involved the movement of a mass containing about 50,000 cu. yd (~38,200 cu. m) of clay (Skempton, 1997).

A cutting excavated at Sudbury Hill in 1900 through weathered London Clay, with slopes as low as 10 degrees, was the site of a slip in 1949 (Skempton, 1964). This reflects the time it takes for instability to become apparent in such overconsolidated clay materials. Following the slip, the only remedial measure that was undertaken was to trim its toe, where it impinged on the railway line.

At Folkestone Warren, the line leading to the port of Dover crossed multiple natural landslides of Chalk overlying Gault Clay. During winter months, high water table levels plus the relaxation of the overconsolidated clay caused movement. Rock falls of chalk onto the slips triggered further landsliding. The construction of a harbour breakwater in 1905 led to beach starvation at the toe of the slope and eventually, in 1915, a catastrophic failure of the Warren occurred (Hutchinson *et al.*, 1980). Remedial measures since this incident have included construction of mass concrete toe weights and drainage adits cut into the slope to improve the effective stress regime.

The Crow's Nest Railway runs through cuttings at the foot of Turtle Mountain in Alberta, Canada. The mountain comprises shales, sandstones and coal seams with an older limestone deposit thrust over the top. The limestone contains joints at right angles to the bedding planes and these are inclined roughly parallel to the glacially oversteepened slopes. In 1903, the crest of the mountain broke off causing 90 million tonnes of rock to slide onto the railway below, killing 70 people (Anderson & Trigg, 1976).

Embankments

Construction begins by removing the top 6 inches (~150 mm) of turf, which is put aside for later sodding down the slopes. All stumps, brush or other obstructions which, by their disturbance of the integrity of the bank or by their decay, might cause sinking or slips would then be removed (Williams, 1850).

The preferred way of building up the main mass of an embankment was then to place thin layers around 2 to 4 ft (~0.6 to 1.2 m) thick, compacted with 'beetles' or punners prior to placing each subsequent layer. Thus the bank would be progressively raised, generally with sufficient time for consolidation to be effected. However, construction of the whole embankment in shallow layers was a slow process. The need for more rapid construction often led instead to the commonly employed method of running out the bank to its full height at one go, by end-tipping (Skempton 1997). Drainage ditches were usually used to prevent water feeding into the core of such embankments, which would otherwise have caused a rapid rise in pore water pressure leading to slippage. A further technique for stabilising an embankment was to set fire to it using coal or timber as fuel. Locally available materials such as clinker or ash would be added to fuse the soil, thereby strengthening the earthworks.

The most common mechanisms of embankment deformation on land owned by the London Underground have been identified as being (McGinnity & Russell, 1996; Chang, 2000):

- unravelling of ash at the crest of an embankment
- shallow-seated localised slips
- deep-seated slope failures

The Hanwell embankment on the Great Western line, 54 ft (~17 m) high, gradually sank as the mass penetrated the fluvio-glacial sands and gravels, alluvium and London Clay strata below. Consequently the ground on either side of the embankment was pushed out and up, and for a distance of 400 ft (~122 m) rose to a height of over 10 ft (~3 m).

In Cheshire, embankments have been affected by subsidence caused by brine extraction associated with salt mining in the area. In particular, the Elton viaduct (which allows the River Wheelock to pass under the railway) has undergone several phases of reconstruction based on the rate of ground failure. When it was constructed in 1842 it was initially a brick arch viaduct. A single span girder bridge replaced this in 1937 (whose abutments could be raised as subsidence progressed). However, in the 1950s it was clear that the structure's design could not adequately cope with the continued subsidence, and the bridge was replaced with the present Elton viaduct (Mountford, 1989). The new design comprised a concrete cellular construction, which takes the form of a traditional embankment with a series of cylindrical pipes to allow the river to flow under the railway. As the structure continues to sink, new layers of pipes may be added to retain stability.

The presence of quick clay under embankments creates another problem. Such materials have low bearing capacity and high rates of settlement. Vibration or shaking of the ground caused by trains can induce liquefaction and cause embankment failure. In 1918, such a collapse occurred near Weesp in the Netherlands, killing 41 people (Turner, 2000).

Permanent way formation

The permanent way may be constructed on anything from a soft compressible material such as peat, to a hard rock. Hence, the variable strength and compressibility of subgrade materials can be problematic.

For the early railways, the importance was not realised of providing a thick layer of hard granular material beneath the sleepers to distribute wheel loads and be freely draining. Hence ash, chalk, burnt clay and other waste materials that were available cheaply and locally were used. It was not until the early twentieth century that the notion of using hard durable aggregate as ballast became universally accepted: gravel, crushed igneous and metamorphic rocks, limestone (not chalk), blast furnace slag and clinker were then used; currently this is being supplied by some 20 quarries across the UK (Smith & Collis, 2001). Engineers responsible for designing the early railways were well aware of the need for good drainage systems but were rarely able to achieve these, resulting in many serious incidents, especially following storms (Cope, 1993).

Cyclic or repeated loading of the subgrade can lead to failure of the permanent way, even when the frequency of the loading is low. Such instability was a particular problem for parts of the Great Western Railway in Oxfordshire (Pryor & Bentley, 1999) where liquefied clay became pumped into the ballast.

Structures

Station buildings

Buildings associated with railway stations in the mid-nineteenth century were built on a scale to reflect the social and economic importance of the railway system. These house the transactions for travel, money collection, book keeping, and baggage handling. The whole business was labour intensive, so there was also the need to accommodate staff, supervisors and management.

The relative lack in revenue subsequently reduced the level of maintenance to the extent that many of these station buildings had become virtually derelict by the mid-twentieth century; demolition became an economic option, but replacement by low cost structures has rapidly led to recurrence of the same problem. The demolition was not a co-ordinated process, more a reaction to the legacy of under-investment in maintenance, and took place within a short space of time; the buildings were close to the operating railway and so much of the work would have had to be carried out under possession. This meant that much of what lay beneath the platform surface still remains.

Particularly troublesome for any new works on or near a demolished station will be the remaining footings of the old buildings. In particular, there may exist the remains of old cellars that have been capped over or filled with material that may have subsequently settled. There may also be old subways and bay platforms. Drainage runs incorporating pre-existing and active land drainage may be buried too. Finally there may be several layers of platform surfacing, becoming covered as the platform levels were renewed and/or raised over the years (The Devil's Guide to the Railway Industry, 2001).

Services that are likely to have been abandoned will include gas and water pipes, for both foul and surface water; the latter were often mixed. At larger stations there may be the remains of large water mains that fed water columns used to replenish the tanks of steam engines.

Hazards include asbestos (often used as lagging for heating and water pipes and as insulation), anthrax spores (present as a result of the importation of horsehair insulation or cow hair used as plaster filler), rotten timber (caused by damp), even dead pigeons (a source of respiratory disease)!

Platforms

The earliest types of platform consisted of low raised areas edged with timber (replaced by paving slabs in the late nineteenth century). They probably had minimal foundations. As carriage designs changed and became less like the stage coaches on which early designs had been based, platforms were built higher and more substantially. There were no national standards as such; each railway company would employ its favoured design. Platforms were built to match the level of the rolling stock, reinforcing the individual company policies.

Platform construction generally comprised a low wall founded on nominal footings. The wall would have been built of brick or stone in the nineteenth century; concrete in the twentieth century. The top of the wall would have been capped with a blue brick having a special nosing coper or with massive stone slabs.

Early platform surfaces were covered with ash, stone slabs or paving, special blue brick paving or a mixture. The fill behind the platform wall generally comprised rubble, mortar and brick-bats. However, the fill was rarely compacted and is therefore susceptible to substantial settlement. As rolling stock became standardised there arose a need to raise many of the old platforms. In many cases this was achieved by a process of placing an extra layer of material on top of the old platform surface. Platforms were also susceptible to subsidence, especially in the Midlands where coal mining occurred.

Goods yards

Goods yard sidings were extensive, covering large areas of land in proximity to stations and industrial premises. Some were devoted to loading, others to the sorting and remarshalling of goods or mineral trains. As business increased so did the scale of the sidings (Williams, 1850).

The construction of sidings and marshalling yards followed a similar pattern to the permanent way, earthworks being utilised to create a sensibly flat topography. As well as the materials used to construct embankments by spreading out waste materials from cuttings, it is likely that a number of waste hot spots are likely to be present, for instance arising from spillage of oils, tars, paints, and minerals, thereby creating a potential health hazard if such sites are to be redeveloped.

Motive power depots

Depots to house locomotives and rolling stock, to service and maintain them, are integral to the railway infrastructure. Steam locomotives were considerably less efficient than diesel or electric, and required a significantly greater number of installations, most of which have now been abandoned. Locomotives and rolling stock standing idle for significant periods will inevitably leak lubricant oils and grease, polluting the subgrade through infiltration. Steam locomotive motive power depots would also have had facilities for removing ash and clinker, usually to pits beneath the track from where the waste could be removed to stockpile. Leachates are likely to have been generated as rainwater trickled through. Diesel motive power depots have the additional risk of fuel oil spillage, and these and electric motive power depots would be handling battery and de-icing fluids, again with risks of spillage and pollution.

Maintenance and renovation of locomotives and rolling stock will be associated with metal work and painting, both of which can lead to introduction of metals (especially heavy metals) to the vicinity.

Geohazards

Geohazards concern the interaction of human activities with natural processes. These have the potential to adversely affect both development and redevelopment. They are caused either by past activity on the land or inherent weaknesses in the fabric of the ground. Those associated with derelict land tend to be small-scale, localised, pervasive in nature and occur at different rates (McCall, 1998). Past research of geohazards has tended to be dominated by the chemical and biological problems of 'contaminated' land with little consideration paid to the physical aspects.

The importance of physical geohazards (e.g. subsidence, heave, settlement, erosion, slope instability, fault reactivation, dissolution, groundwater and underground gases) has tended to be understated even though major economic and financial losses have occurred. One of the centres to initiate geohazard studies was the Building Research Establishment (BRE), notably work on the effects of building on artificially-emplaced fill in opencast mines, quarries and landfills (e.g. Skinner & Charles, 1999; Watts & Charles, 1999). There is a general lack of published material concerning other types of artificial land and associated problems (Charles, 1999).

However, accidents can nevertheless occur. Some are caused by the ground (geohazards), others by human error, defective materials, vandalism or mechanical failure. Fire, water, subsidence and rock falls are the main hazards known to cause railway accidents. In the days of steam, sparks, red-hot ashes, oil, or naked flames coming from the locomotives could ignite fires. The prevalence of coal provided the fuel to spread the fire, most notably in embankments that had been constructed with ash or clinker (light and readily available, thus physically suitable for repairing embankments that had settled). Temporarily the railway would have to be closed, but long-term benefit could arise if the heat fused the clay minerals, effectively creating *in situ* brick that reinforced the earthworks.

Examples of incidents caused by geohazards

At Vriog, near Fairbourne, on the former Cambrian section of the Great Western Railway (GWR), where the track ran on a shelf on the cliff side, a landslide hit the first train of the day in 1933. Fortunately, only the engine was swept over the edge. The accident investigation report pointed out that a similar incident had occurred nearby in 1883. Therefore, the 1933 incident should have been preventable with knowledge and mitigation of the geohazard. Following the 1933 accident, the GWR built an avalanche shelter over the line to protect future trains (Hoole, 1983).

In 1909, a train derailment occurred near Skinnigrove caused by subsidence associated with ironstone mining beneath the viaduct on which the railway ran (Hoole, 1983). It appears that the geological records had not been checked for this potential geohazard. Two years after the incident, local mining spoil was

used to convert the viaduct into an embankment as the proximity of underground mining had made the piers unsafe.

In October 2001, two large landslides closed the line to Kyle of Lochalsh near Stromferry in northwest Scotland. The failures were triggered by prolonged heavy rainfall. Both were at the toe of steep hillsides several hundred metres high, and the debris from one of them struck a passenger train travelling between Kyle of Lochalsh and Inverness (Anon., 2002). Detailed site assessment indicated that the best solution was to realign the railway. This involved building a 400 m long, 7 m high causeway on the tidal shore of Loch Carron for the new section of track to run on. A 15 m high buttress was built at the toe of the hillside to protect the track from further slides. The project cost £2M, placing 50,000 tonnes of rockfill from quarries on the Isle of Skye and Dingwall, together with more than 10,000 tonnes of armour stone from Oban and Inverness for coastal protection.

Other geohazards that impact on railways are heat (during excavation of tunnels), earthquakes and volcanic activity. Tunnelling through the Alps had to overcome the presence of very hot groundwater, e.g. the Swiss Simplon tunnel in 1905 encountered temperatures of around 55 °C. These were conquered either by releasing the water pressure in the fissures using explosives or by pumping cold water under high pressure into the fissures thereby diluting them down to a temperature the workers could stand (Fox, 1914). More recent solutions include geomembranes and sealants.

Earthquakes may cause severe distortion or even failure of permanent way in seismically active areas (e.g. USA, Japan, Greece, India). However, development of flexible materials and designs to absorb seismic energy are reducing the severity of such disturbance.

The threat of volcanic activity is spatially limited, but an ever present geohazard for the people that live in the vicinity of volcanoes such as Mount Etna in Sicily. Eruptions may occur intermittently, but disruption of services is inevitable when they occur. The circum-Etna railway has had to be re-routed on several occasions as the direct result of new lava flows.

Overview of geohazards guidance

In the UK, guidance for dealing with geohazards has until now tended to have been considered separately from the planning and geotechnical perspectives. Advice for planners is dominated by the Planning Policy Guidance (PPG) and Mineral Policy Guidance (MPG) notes published by the UK Government. However, the majority of these documents rarely explicitly consider the impact of geohazards on a site, except where potential development would be on unstable land (PPG 14), coastal sites (PPG 20) or on areas susceptible to flooding (PPG 25). PPG 14 has recently been amended to include more detailed guidance on how to deal with landslides and subsidence. The MPGs deal with instability at current or former mining areas.

Geotechnical guidance deals with geohazards in a more general way, published information being contained within:

- British Standards Codes of Practice, BS 5930:1999 for site investigations and BS 10175:2001 for investigation of potentially contaminated sites
- the Institution of Civil Engineers Working Party report, "Managing Geotechnical Risk" (Clayton, 2001), sponsored by the ICE and the Department of Transport, Local Government and the Regions
- the Construction Industry Research and Information Association (CIRIA) report, "Remedial Treatment for Contaminated Land (Volume III) – Site Investigation and Assessment".

The two British Standards (BS 5930:1999 and BS 10175:2001) focus on physical features and contamination issues respectively, with little overlap between them. The Total Geological Model (Fookes *et al.*, 2000) describes a currently evolving methodology for depicting site investigation information, while the two professional reports describe techniques for managing geotechnical risk associated with geohazards.

The communication of geohazards and risk to the non-geoscientific professions and the wider public forms a major focus for the recently established Geohazards Working Party (under the auspices of the Geological Society of London) and is due to report at the 10th Congress of the International Association of Engineering Geology and the Environment in 2006.

The consequences of railway construction for geohazards

Ground materials are subject to a range of processes that affect their composition, their arrangement, their geotechnical properties and their interactions with fluids, gases and other materials within the ground. Temporal and spatial variations will determine the likelihood of geohazards on any specific site. The potential occurrence of geohazards presents one of the main uncertainties for the investigation of derelict railway land.

The factors that control the potential occurrence of geohazards on railway land can be considered under three themes with regard to their potential impact on development:

- *site history* – the transportation of materials that constitute the site, their age, provenance and mode of construction
- *material processes* – the internal dynamics of each material and with each other (at the macro and micro scale, as well as fluids and gases *in situ*)
- *boundary conditions* – the interaction of materials across interfaces. Key components at the macro scale are the lateral extent of each material and the ground profile.

These themes apply equally to land once used for railway purposes. They may be characterised with respect to five geological principles:

- stratigraphy
- geomorphology
- hydrogeology
- land use history
- industrial archaeology

Stratigraphy

Stratigraphy is the branch of geology concerned with the characteristics and attributes of rocks that occur in strata, and the interpretation of such strata in terms of derivation and geological background (Lapidus, 1990). Traditionally, stratigraphy has been used as a tool to establish the Earth's geological history from the sequence of events that have occurred through time.

As the railways were built, material was removed (eroded) from cuttings and placed (deposited) in embankments and the permanent way by man (process). Embankments and track bed were commonly built up using local materials, preferably from cutting excess. If materials were in short supply, resources from further afield would have to be imported, but at greater cost. Mixed sources could thus comprise both local resources and materials from further afield. In essence, railway construction provides the basis for an artificial geological stratigraphy.

The character of the near surface natural materials on which the permanent way stands (largely Quaternary deposits) is a vital consideration. Such deposits (e.g. alluvium, till, loess, peat, fluvio-glacial sands and gravels) would have been subjected to various events and processes in recent geological time, for instance: freezing and thawing, sea level change, groundwater movement. The boundary between Quaternary deposits and man-made materials is therefore crucial to understanding the geotechnical character of former railway land.

Geomorphology

Geomorphology is the study of surface features of the Earth, especially landforms and the chemical, physical, and biological factors that act upon them (Lapidus, 1990). The geomorphological setting of a site is significant with regard to the activity of near-surface processes, hydrogeological, periglacial, fluvial, and marine erosional and depositional features (Hutchinson, 2001).

River action removes material from one place and deposits it elsewhere. Surface waters can downcut rapidly through the strata creating potential ground problems with very steep slopes. Coastal erosion includes wave action on cliffs which causes slope instability and eventual collapse. Beach protection measures (e.g. groynes) for one part of a coastline can lead to instability problems at another, as the material that would normally be deposited is prevented from doing so and erosion takes place instead. This was a particular problem for Folkestone Warren where waves were able to actively erode the toe of the slope.

Solifluction is the downslope movement of water-saturated debris above frozen ground and can involve movement of up to 1 metre per year, on slopes as low as two degrees. Head deposits are unsorted debris resulting from solifluction and are entirely comprised of local upslope material. Soil creep is the slow movement of a deposit due to gravity, aided by surface water, vegetation and burrowing activity. Materials susceptible to creep (e.g. soil) can take many years, even decades, to settle and become stable.

Hydrogeology

Pore water pressure depends upon the groundwater regime. Excess pore water pressure leads to physical instability of the ground. The groundwater chemistry can dissolve material and also lead to growth of new materials, *e.g.* iron pyrites decomposes upon exposure to water and air, soluble materials dissolve leaving voids, and produce acid solutions that can attack metalwork; the corrosion of the cast iron lining on the Northern Line of the London Underground at Old Street provides a classic example (Rainey & Rosenbaum, 1989).

Flood events, of course, are also hazardous. The presence of a pool of water on a site may be indicative of poor drainage that may influence future usage of the site. Ground instability is also associated with groundwater emergence (from springs and by seepage), perched aquifers and artesian flow.

Land use history

Land use history can be deduced by examining topographic maps such as those compiled by the Ordnance Survey, from which a detailed picture of the history of a site can be built up. Their coverage, especially at 1:10 000 (or the older 6 inches to 1 mile) scale, is generally good and the first editions are available on the Internet at www.old-maps.co.uk. Land that may appear natural may in fact be modified by humans and thus exhibit different characteristics to natural ground. For instance, excess material removed from cuttings may have been spread out on land adjacent to the railway if it wasn't required or couldn't be used in embankments nearby.

Indications of past activity are not always clear. In some cases, evidence may have been swept away in the course of subsequent development. Field observation is an essential prerequisite to indicate how and why the ground was developed as it has, the types of activities and processes likely to have acted on it and any subsequent disturbance.

Industrial archaeology

Industrial archaeology is the science which deduces a knowledge of past times from a study of their existing remains (Nock, 1981). The principle relevant to this paper is the exploration of the remains (evidence) of derelict railway land.

This incorporates the inspection of the raw materials used to construct the railways, assessment of where they came from, consideration of any special techniques/processes used in their construction, observation/analysis of the composition/structure of the material, and any change that has affected it since placement.

Railway land in Nottingham

The character of railway land can be illustrated by examining the situation in the city of Nottingham. Four scenarios typify the range of geotechnical characteristics in terms of Nottingham's geology, land-use history and groundwater conditions.

- Early railway construction – the floodplains of the Trent
- Later construction – The Great Central Railway
- Redevelopment – Colwick Loop (Netherfield)

Early railway construction – the floodplains of the Trent

The first railways benefited from breaking new ground so there were few physical constraints regarding where they could be routed. Thus early railways tended to follow the courses of rivers, where the land was flatter. The Midland Railway along the Trent floodplain and the Great Northern Railway in the Leen valley reflect this in Nottingham. The main problem for the pioneering railways occurred when landowners refused to let a railway company cross their land, which could result in detours over 'less suitable' ground.

The Midland Counties Railway (MCR) constructed lines from Derby to Nottingham and from Long Eaton to Rugby (to connect with the London & Birmingham Railway) in 1832. Six years later, the whole line was under contract and the Nottingham station house, engine-house and goods shed were built on what had previously been meadow land to the River Trent.

In 1844, the MCR amalgamated with the North Midland Railway and the Birmingham & Derby Junction Railway to form the Midland Railway (MR). A year later, the Derby to Nottingham line was extended, via a junction over Wilford Road, to Lincoln. However, the extra traffic led to the station becoming increasingly constricted so a new station was built in 1848, at the east end of Station Street. This station remained in use until 1904 when the present station opened (facing west) onto Carrington Street (Wilson, 1996).

As the MR expanded, a viaduct had to be constructed over Queen's Road, in 1869. The completion of this viaduct coincided with other alterations and improvements to the station, including construction of a third platform, new goods lines and infilling of the Westcroft branch canal, constructed in 1839. The presence of natural and man-made materials indicates that, even in these early days, various phases of development and redevelopment caused problems to the ground characteristics.

Later construction – The Great Central Railway

The last mainline railway company to drive a line through Nottingham was the Great Central Railway (GCR), formerly called the Manchester, Sheffield & Lincolnshire Railway. It made plans to build a line to enable swift travel to London and the continent, and opened in 1899. The GCR route bisected the city using a north – south route in an almost straight line (Forster & Taylor, 1991).

Cuttings were generally excavated through moderately strong rock (e.g. Sherwood Sandstone, Magnesian Limestone) and could be thus almost vertical in nature. The cutting spoil was used to create the embankments beyond. Indeed, a good balance was achieved between materials removed and material used to form earthworks.

The route was a prime example of the Victorian attention to detail. It crossed the Leen valley at Bulwell on an enormous blue brick viaduct, and was then carried over an embankment on the approach to New Basford before plunging into the depths of the Sherwood Rise and Mansfield Road tunnels, to emerge at Victoria Station in a deep sandstone cutting. It then entered a cut and cover tunnel (through the Sherwood Sandstone) under Thurland Street before reappearing at Weekday Cross and crossing over the Midland Station and the River Trent on a succession of bridges and viaducts (Wilson, 1996).

However, the GCR route closed in the 1960s following the rationalisation of the national network. Much of the route remains derelict to this day, although the Victoria Station (with its famous clock tower) now houses the Victoria Shopping Centre and some of the bridgework over Canal Street is being incorporated within part of the NET tram route.

Redevelopment – Colwick Loop (Netherfield)

Netherfield was once industrialised and contained extensive sidings, locomotive sheds (steam and later diesel multiple units), as well as sidings to various factories, and warehouses dating back to 1878 (Forster & Taylor, 1991). To the west was terraced housing and to the east were green areas, infilled brick clay pits, gravel pits, and a coal mine. Today, the land-use has changed significantly: an industrial estate and playing fields have been built over Colwick sidings, and low-rise residential buildings dominate Netherfield (Henshaw, 1999).

The site lies on the flood plain of the River Trent and had to be raised to avoid flood inundation (to about 20 m AOD). The Quaternary stratigraphy comprises alluvium and lacustrine deposits, backed by low river terraces. Thus the risk of adverse groundwater conditions was clear from the geology. However, the frequent flooding of the River Trent in historical times meant substantial provision was required from the outset.

The fill for these sites largely consists of local waste materials. Field investigation of the playing fields area revealed that Colwick sidings comprised the following materials:

- smelting waste (clinker and unburnt charcoal)
- bricks
- sandstone rubble
- glass bottles.

The style of the glass bottles effectively acts as a stratigraphical marker, revealing that they were dumped around the early twentieth century. These, and the above materials, were mixed with a sandy reddish-brown soil. The slopes of the embankments were poorly vegetated with pits and hollows signifying instability. Kinks in some of the tree trunks as they grew also suggest slope stability problems. The uneven nature of the surface of the playing fields also indicates that settlement is still occurring.

The industrialised area is more difficult to inspect due to the many phases of redevelopment the area has undergone. The top half metre or so of the site contains demolition rubble (e.g. bricks, concrete fragments, stone tiles, plasterboard, metal wire) as well as materials from fly-tipping (waste paper, metal gas canisters). It can be assumed that the underlying material is similar to that of the playing fields. The area is relatively poorly vegetated with sparse trees and short grasses. Pits and holes are common. The fact that redevelopment consists of low-rise buildings suggests settlement had not stopped at the time of construction.

Today, railways are still a prominent feature within Nottingham. The southern fork of the Cinderhill Colliery branch is currently being redeveloped as part of the Nottingham Express Transit (NET) scheme, while the northern (GNR) fork remains derelict (with its imposing embankment). The remaining land is currently used for housing and light industry.

The derelict GNR embankment still retains a number of railway features. Remains include a platform (sandstone wall, concrete slabs/fence posts), sleepers and associated brick (black) structures. The sandstone wall is cracked indicating differential settlement has occurred. The embankment consists of a core of Sherwood Sandstone blocks (obtained from local cuttings) covered by a dark organic rich soil in some places and a clinker-rich man-made material in other places. As well as clinker, this artificial layer contains fused bits of coal shale and sandstone, man-made clinker and slag (indicated by the presence of air bubbles) and hardened globules of a tarry substance. It is likely that this man-made deposit came from local metal foundries and was used to repair past instability in the embankment.

Conclusions

The legacy of railway construction significantly impacts on the development of urban space. The rapid expansion of railways in the UK during the nineteenth century has created characteristic geotechnical conditions but contain a number of potential geohazards. Nevertheless these may, to a substantial extent, be anticipated by evaluation of historical development and the railway context.

Particular ground conditions are susceptible to specific geohazards. Of the five themes presented, knowledge of the stratigraphy is paramount; the method of construction, types of materials used and where the materials came from are also important considerations. Understanding the site geomorphology and the hydrogeological regime forms the basis for identifying the internal material processes, how they interact and what is likely to be happening at material boundaries. Industrial archaeology, interfacing with land use history provides the context for tying in the evidence on the condition of the ground, so providing the basis for effective evaluation of site conditions as a precursor to successful redevelopment and regeneration of derelict land.

Acknowledgements

The authors would like to thank their colleagues at the Nottingham Trent University and the British Geological Survey for their support in developing the ideas presented in this paper, in particular Dr. Ron Blake for his guidance with the Nottingham case studies at Colwick and Basford.

References

Anderson, J.G.C. & Trigg, C.F., 1976. *Case-histories in engineering geology*. Elek Science, London. 204 pp.

Anon., 2002. Landslip repair should see Kyle Line reopen for Easter. *Ground Engineering* (February), p 6.

Appleton, J.H. & Appleton, R.J., 1970. *Disused railways in the countryside of England and Wales*. HMSO, London. 82 pp.

British Standards Institution, 1999. *Code of practice for site investigations*. BSI, London, BS 5930. 206 pp.

British Standards Institution, 2001. *Investigation of potentially contaminated sites – Code of practice*. BSI, London, BS 10175. 76 pp.

Chang, J., 2000. Stabilisation of railway embankments. *In:* British Geotechnical Society. *Proceedings of the Sixth Young Geotechnical Engineers Symposium, 12–14 April, City University, London*. 71–72.

Charles, J.A., 1999. *Regarding published research of artificial land* (pers. comm.).

Clayton, C.R.I., 2001. *Managing Geotechnical Risk*. Institution of Civil Engineers and Thomas Telford. 80 pp.

Cope, G.H. (ed.), 1993. *British railway track – design, construction and maintenance*. The Permanent Way Institution. 628 pp.

CTRL, 2003. *Channel Tunnel Rail Link: Press statement regarding landslide in Lavender Street, E15*. http://www.ctrl.co.uk/press/contract.asp?ID=3668 [Accessed 20 February 2003].

Department of the Environment, Transport and the Regions, 1998a. *Planning for the Communities of the Future (Guidance Report)* [online]. http://www.planning.odpm.gov.uk/future/index.htm [Accessed 28 Mar 03].

Department of the Environment, Transport and the Regions, 1998b. *A New Deal For Transport: Better For Everyone (White Paper).* http://www.dft.gov.uk/itwp/paper/chapter4/6.htm [Accessed 28 Mar 03].

Department of the Environment, Transport and the Regions, 1999. *Towards An Urban Renaissance (Urban Task Force Report).* http://www.urban.odpm.gov.uk/whitepaper/taskforce/renais/index.htm [Accessed 28 March 2003].

Department of the Environment, Transport and the Regions, 2000. *Our Towns and Cities: The Future – Delivering and Urban Renaissance (White Paper).* http://www.urban.odpm.gov.uk/whitepaper/ourtowns/fulltext/index.htm [Accessed 28 March 2003].

Fookes, P.G., Baynes, F.J. & Hutchinson, J.N., 2000. *Total Geological History: A model approach to the anticipation, observation and understanding of site conditions.* Proceedings for GeoEng 2000: the IAEG/ISRM/ISSMGE International Conference on Geotechnical & Geological Engineering (held in Melbourne 19–24 Nov 00), Technomic Publishing Co., Lancaster, Pa, USA, **1**, 370–460.

Forster, V. & Taylor, B., 1991. *Railways in and around Nottingham – scenes from the past, no. 11*, Foxline Publishing, Stockport. 98 pp.

Fox, F., 1914. How the Swiss Built the Greatest Tunnel in the World, 1905. *In:* Tappan, E.M. (ed.). *The World's Story: A History of the World in Story, Song and Art, Vol. VII: Germany, The Netherlands, and Switzerland*, Boston: Houghton Mifflin, pp 601–605. http://www.fordham.edu/halsall/mod/1905simplon.html [Access 17 Mar 03]

Henshaw, A., 1999. *The Great Northern Railway in the East Midlands: Colwick Yards – Nottingham London Road – Gedling – Basford.* The Railway Correspondence & Travel Society, Nottingham, 108 pp.

Hoole, K., 1983. *Trains in trouble: Railway accidents in pictures (Volume 4).* Atlantic Transport & Historical Publishers, Cornwall. 68 pp.

Hutchinson, J.N., Bromhead, E.N. & Lupini, J.F., 1980. Additional observations on the Folkestone Warren landslides. *Quarterly Journal of Engineering Geology*, **13**, 1–31.

Hutchinson, J.N., 2001. Reading the ground: morphology and geology in site appraisal – Fourth Glossop Lecture. *Quarterly Journal of Engineering Geology & Hydrogeology*, **34**, 7–50.

Lapidus, D.F., 1990. *Collins dictionary of geology*. Harper Collins. 565 pp.

McCall, G.J.H., 1998. Geohazards in the urban environment. *In:* Maund, J.G. & Eddleston, M. (eds). *Geohazards in Engineering Geology*. Geological Society, London, Engineering Geology Special Publications, **15**, 309–318.

McGinnity, B.T. & Russell, D., 1996. Investigation of London Underground earth structures. *In:* Craig, C. (ed). *Advances in Site Investigation Practice*: Proceedings of the Institution of Civil Engineers, London, 230–242.

Mountford, J.B., 1989. *Cheshire salt and subsidence*. M.Sc. thesis, Imperial College of Science and Technology, London. 180 pp. (unpublished).

Nock, O.S., 1981. *Railway Archaeology*. Patrick Stephens, Cambridge. 192 pp.

Phipps, P.J. & McGinnity, B.T., 2001. Classification and stability assessment for chalk cuttings: the Metropolitan Line case study. *Quarterly Journal of Engineering Geology and Hydrogeology*, **34**, 353–370.

Pryor J. & Bentley S.P., 1999. Sources of historical information on the condition of railway embankments, *Ground Engineering* (September), 22-23

Rainey, T.P. & Rosenbaum, M.S., 1989. The adverse influence of geology and groundwater on the behaviour of London Underground railway tunnels near Old Street station. *Proceedings of the Geologists' Association*, **100**, 123–134.

Skempton, A.W., 1964. Fourth Rankine Lecture: Long-term stability of clay slopes. *Geotechnique*, Institution of Civil Engineers, London, **14**, 77–102.

Skempton, A.W., 1997. Embankments and cuttings on the early railways. In: Chrimes, M. (ed.). *Studies in the History of Civil Engineering - The Civil Engineering of Canals and Railways before 1850*, **7**, 291–307.

Skinner, H.D. & Charles, J.A., 1999. Problems associated with building on a variable depth of fill. *Ground Engineering* (July), 32–35.

Smith, M.R. & Collis, L. (eds), 2001. *Aggregates: sand, gravel and crushed rock aggregates for construction purposes*. Engineering Geology Special Publication, Geological Society, London, **17**, 285–289.

The Devil's Guide to the Railway Industry, 2001. *The Railway Industry*. http://www.devilsguide.com/home.asp [Accessed on 22 November 2001].

Turner, A.K., 2000. The impact of geology on society. *Proceedings of an inaugural lecture, Delft University of Technology*. p 23.

Watts, K.S. & Charles, J.A., 1999. Settlement characteristics of landfill wastes. *Proceedings of the Institution of Civil Engineers: Geotechnical Engineering*, **137**, 225–233.

Williams, F.S., 1850. *Our Iron Roads: Their History, Construction and Administration (Volumes One & Two)*. Reprinted in 1981, Gresham Books Ltd, England. 511 pp.

Wilson, J.P., 1996. *The Development of Nottingham's Railways*. Nottingham Civic Society. 32 pp.